Designing Engineering and Technology Curricula

Embedding Educational Philosophy

Synthesis Lectures on Engineering, Science, and Technology

Each book in the series is written by a well known expert in the field. Most titles cover subjects such as professional development, education, and study skills, as well as basic introductory undergraduate material and other topics appropriate for a broader and less technical audience. In addition, the series includes several titles written on very specific topics not covered elsewhere in the Synthesis Digital Library.

Designing Engineering and Technology Curricula: Embedding Educational Philosophy
John Heywood
2021

Introduction to Engineering Design
Ann Saterbak and Matthew Wettergreen
2021

Visualizing Dynamic Systems: Volumetric and Holographic Display
Mojgan M. Haghanikar
2021

Biologically Inspired Desgin: A Primer
Torben A. Lenau and Akhlesh Lakhtakia
2021

Engineering Design: An Organic Approach to Solving Complex Problems in the Modern World
George D. Catalano and Karen C. Catalano
2020

Integrated Process Design and Operational Optimization via Multiparametric Programming
Baris Burnak, Nikolaos A. Diangelakis, and Efstratios N. Pistikopoulos
2020

Project Management for Engineering Design
Charles Lessard and Joseph Lessard
2007

Relativistic Flight Mechanics and Space Travel
Richard F. Tinder
2006

Designing Engineering and Technology Curricula: Embedding Educational Philosophy

John Heywood

ISBN: 978-3-031-03752-8   paperback
ISBN: 978-3-031-03762-7   PDF
ISBN: 978-3-031-03772-6   hardcover

DOI 10.1007/978-3-031-03762-7

A Publication in the Springer series
SYNTHESIS LECTURES ON ENGINEERING, SCIENCE, AND TECHNOLOGY

Lecture #17
Series ISSN
Print 2690-0300   Electronic 2690-0327

Designing
Engineering and Technology
Curricula
Embedding Educational Philosophy

John Heywood
Trinity College Dublin–University of Dublin

SYNTHESIS LECTURES ON ENGINEERING, SCIENCE, AND TECHNOLOGY #17

ABSTRACT

The intention of this book is to demonstrate that curriculum design is a profoundly philosophical exercise that stems from perceptions of the mission of higher education. Since the curriculum is the formal mechanism through which intended aims are achieved, philosophy has a profound role to play in the determination of aims.

It is argued that the curriculum is far more than a list of subjects and syllabi, or that it is the addition, and subtraction, of items from a syllabus, or whether this subject should be added and that subject taken away.

This book explores how curricular aims and objectives are developed by re-examining the curriculum of higher education and how it is structured in the light of its increasing costs, rapidly changing technology, and the utilitarian philosophy that currently governs the direction of higher education.

It is concluded that higher education should be a preparation for and continuing support for life and work, a consequence of which is that it has to equip graduates with skill in independent learning (and its planning), and reflective practice. A transdisciplinary curriculum with technology at its core is deduced that serves the four realities of the person, the job, technology, and society.

KEYWORDS

ABET, action research, adaptability (-adaptive behaviour), affective, AI, aims(s) (-of education), alignment, assessment, attitudes, augmented reality, automation, beliefs, bridging concept, cognate work (-studies), cognitive, communicate (-ion), community, competence (-professional), concepts, conflict, content, constructivism, cross-domain transfer, curriculum (-drift, received), design (-of curriculum), development (cognitive/affective), disciplines, e-learning, emotion(s), employment, engineering, engineering education, engineering literacy, entering characteristics, epistemology, ethics examination(s) (-boards), fees (-tuition), flexibility, forecasting, general education, generalisation (-stage of), higher education, identity, ideologies (learning centred, scholar academic, social efficiency, social reconstruction, instruction, intelligence (-academic, practical), jobs, key concepts, knowledge (-circle of, as information, as design, strategic), labour arena, language(s), learning (-by doing, depth, independent, organization, surface), liberal education, linguistic analysis, literacy (-engineering, technological), logic, logical positivism, meaning, middle skill jobs, motivation, non-cognate work, objectives, object worlds, occupation(s), occupational transfer gap, outcomes, ontology (-ical), over-arching concept, overload (-curriculum), philosophy, phronesis, power, precision (-stage of) principles, problem-based learning, profession(s) (-al), project-based learning, quality (-teaching), realism, reflection (reflective capacity), relationships, robotics

For Sue Kemnitzer

Contents

Preface

The first purpose of this book is to show that the curriculum is far more than a list of subjects and syllabi, or that curriculum design is the addition and subtraction of items from a syllabus, or whether or not this subject should be added and that subject taken away. The second purpose is to demonstrate that the design and development of the curriculum is a profoundly philosophical exercise that stems from perceptions of the mission of education, in particular higher education. Since higher education is a sub-system, it cannot be separated from what happens to graduates in the future, or to what has happened to them in the past, or for that matter, its responsibility in that development. Therefore, the determination of its aims is an exercise of the utmost significance. Since the curriculum is the formal mechanism through which intended aims are achieved, philosophy has a profound role to play in their determination (2.3). The significance of the curriculum process is illustrated through the consideration of the problem of "curriculum overload" (2.4).

To illustrate the role of philosophy in the curriculum process, attention is drawn to its demand that the terminology used should be clear and, at the same time, it should help to clarify it (1.1). In spite of this there is much ambiguity in educational thinking even in statements of objectives (outcomes) (1.3), and there are huge difficulties in defining subject areas like engineering and technological literacy which are the focus of this text (1.4). Matters are complicated by the fact that engineers have to speak different "specialist" languages to different people in the process of design and manufacture (1.5), notwithstanding globalisation, and the possibility of having to communicate in foreign languages.

However, Joseph Pitt reminds us that "philosophy is not all about how language works, but how people use it to communicate better and solve problems affecting their immediate lives. It is not all about logic, but also about how people reason informally and why that is, in the long run, more important than producing a consistency theorem. It is not just about the structural flaws in various ethical theories, but about what kind of a world we should be building in order truly to live the good life. In short, it is about people. It is about how all of us relate to each other and to the world. And if it is about all of us, then it necessarily includes engineers."

A model of the curriculum process is presented in Chapter 2 (2.3). It illustrates the need for alignment between aims and objectives, learning, assessment, and the learning characteristics of the students. Attention is drawn to the influence of accreditation and quality assurance agencies on the process. This process is illustrated with the example of the overloaded curriculum (2.4)

Chapter 2 begins by defining operational philosophy as the value system that drives a particular curriculum, syllabus, course, or teaching session. Because the beliefs we have about education determine how we think students learn and how they should be taught, they should always be subject to scrutiny, philosophically and evidentially. For example, it seems that current responses to engineering needs as expressed in the curriculum stem from a mix of ideologies (2.5). Nevertheless, many engineering educators still see the purpose of the curriculum as information giving.

However, we do need to be able to defend our aims and, at the same time, come to an agreed statement of purpose. In such debates many aims will be proposed and in order to obtain consistency they need to be screened for contradictions, while and at the same time create a small but significant list of aims (2.6). Philosophy provides the value system against which the aims may be evaluated. Ralph Tyler argued that this exercise had to be followed by a second screen using instructional psychology and models of cognitive and personal development should be added (2.6). It is argued here that these screening processes do not follow a linear succession, but take place in parallel, and interact with each other. It is further argued that it requires that every participant in the process should have a defensible philosophy and psychology of learning and development.

Given that a curriculum is a specification for action, curriculum design is the process of inquiry that leads to that specification. It is the reverse of the more normal case where a designer is given the specification. The designer considers the external constraints that bear on the curriculum considered as an open system.

The purpose of the chapters that follow (3–6) is to illustrate screening by re-examining the curriculum and structure of higher education in light of (a) the increasing cost of higher education and its effect on tuition, fees, and quality, (b) rapidly changing technology on employment, and (c) the utilitarian philosophy that currently governs the direction of higher education.

Arguments about student loans in both the UK and the U.S. have led to questions about the structure of higher education, the quality of teaching (3.6), the content of the curriculum, the irrelevance of degrees in certain subjects, and, above all, the purpose of higher education (3.1 and 3.4). Traditionally, an undergraduate degree in the UK takes three years, where as in the U.S. it takes four, with students in classes for just over half the year. Yet, these current structures have no real rationale. Therefore, it would be easy to re-organise the academic year so as to compress three years into two and four years into three. In the UK, it was recently estimated that this change would reduce the costs of tuition by 20%, which is why universities are being encouraged to run two year programmes (3.5). In both countries it is difficult to understand why it is that three- and four-year programmes have become the norm except in so far as the general curriculum, and within subjects specifically.

At a more fundamental level, the changes that are taking place lead to questions about the purposes of higher education: "for whom?" and "for what?" (3.3 and 3.8).

Studies of the impact of technology, in particular AI, produce both optimistic and pessimistic scenarios of the future of employment (4). It is clear that, at all levels, roles will change. It is also clear that some jobs will be lost and not replaced in geographical areas where they were lost. New types of middle skill work will be created (4.5), and many new tasks will be hybrid (4.6). In the long run, technology will profoundly influence our understanding of what it is to be a professional (4.4). As Jaron Lanier pointed out in *"Who Owns the Future"* the choices that we make about how we control technology are of profound significance, especially for those who will follow us (5.1).

Technological and organizational displacements will force some workers (at all levels) to seek employment in fields that are non-cognate with the fields they are currently in. It is expected that everyone will have to become more adaptable than they have been in the past in order for them to make the career changes life may demand. More specifically, they will require skill in non-cognate or cross-domain transfer. This will require of individuals not only to have the motivation to transfer across domains but skill in devising programmes of learning that will enable them to move forward with their careers. A basic aim of higher education is therefore to equip graduates with skill in independent learning, and the associated quality of reflective practice (7). Such skill will enable the graduate to gain new specializations and expertise in a system of lifelong learning that is integral to the needs of personal and professional development rather than an addendum (6).

Chapter 5 examines the criticisms of newly qualified engineering graduates in particular and higher education in general that have been made by industrialists (5.1). Examination of these complaints show that they relate as much to the affective domain as they do to cognitive. They have been called "soft skills" but now many people use the term "professional skills". They have also been called "personal transferable skills". Some responses of policy makers are described. Comparing these lists with the views of laypeople and experts on intelligence categorised by R. J. Sternberg shows that many of the qualities demanded are very much those that liberal education educators claim to develop (5.3). If there is to be change it will have to have an adequate philosophical framework. The need for lifelong learning is again stressed

The utilitarian model that governs higher education in both UK and the U.S. is beginning to be questioned (5.3). In focusing on the perceived economic value of STEM degrees at the expense of the arts and humanities both the person and society have been neglected. An aim of education is, therefore, to develop a balanced curriculum that takes into account the economy, society, work and the person. In any case, criticisms of current programmes by industrialists are primarily concerned with the lack of development of "professional" or "soft" skills which belong primarily to the affective domain, and these can only be developed by the inclusion of concepts from the arts, humanities, and social science. Thus, solving complex engineering problems contributes to a person's liberal education such as that conceived by John Henry Newman, which necessarily involves the cognitive and affective domains of human development.

The other criticism of industrialists that graduates are not able to work immediately in industrial situations is shown to be unrealistic since many of the competencies required by industry are specific to the context in which they are required, and therefore, require support from industry for their development (5.4).

Given the significance of adaptability and its importance to cross-domain transfer, the whole of Chapter 6 is devoted to this topic. Organizations that perceive themselves to be learning systems will necessarily organize themselves as a community of learners (6.2). Such communities should enhance the development of the individual (6.5). The neglect of the personal in higher education becomes apparent (6.5). The ability for cross-domain transfer is enhanced if a person has had a broad-based education that enlarges the mind on which they can build the specialist studies they will need from time to time (6.3; 6.4).

Chapters 3–6 show how screening using philosophy and psychology side-by-side to derive aims and objectives links them to the type of curriculum that is required. Chapter 7 describes a model of a two-year programme to meet the requirements of a "basic" higher education for preparation for further learning, living, and working in an advanced technological society. It begins with a review of the preceding chapters in terms of the aim of higher education as a preparation for, and continuing support for, life and work. One consequence is that it has to equip graduates with skill in independent learning (and its planning), and the associated quality of reflective practice (7.1).

All of us have to face a world that is considerably challenged. Solutions to its problems will invariably require information from many different knowledge areas (7.2) but they will not be amenable to a subject-based approach to the curriculum. A transdisciplinary approach to the curriculum based on A. N. Whitehead's theory of rhythm in learning (7.5) focused on key concepts (7.3, 7.4), and problem-/project-based learning is proposed to run in parallel with a programme in technological literacy. This curriculum serves as a framework for the four realities of the person, the job, technology, and society. Without such a programme the conflicts inherent in the concept of technological literacy cannot be resolved.

John Heywood
November 2021

Acknowledgments

The ideas for this text were originally conceived for a distinguished lecture at the American Society for Engineering Education in 2012. It was subsequently revised and published in the *Cambridge Handbook of Engineering Education Research* under the title "Engineering at the Crossroads; Implications for Educational Policy Makers". A paper titled "The Response of Higher Education to Changing Patterns of Employment" was also given at the same conference. Since then, several papers have been presented at these conferences, but of more significance has been a continuing discussion among my friends and colleagues about how these ideas might be developed.

I am deeply indebted to Alan Cheville, Charles Larkin, and Mani Mina who, in talking with me almost weekly, helped me re-shape and develop my thinking. Charles has offered a substantial critique, and Mani has been through the text word-for-word, but any errors are mine.

Elizabeth Pluskwik has also been through the script and advised on what to leave in and what to leave out, what is difficult, and what is easy. Thank you. Again, thank you to Arnold Pears for involving me in the exercise for which the script was written.

Over the years many individuals have contributed to my thinking and supported my efforts. I would like to mention Dick Culver, Carl Hilgarth, John Krupczak, Russell Korte, and Karl Smith in the U.S., Bill Grimson and Michael Murray in Ireland, John Cowan in Scotland, Glyn Price, and the late George Carter and Deryk Kelly in England

John Heywood
November 2021

CHAPTER 1

The Languages We Speak

1.1   THE MEANING OF WORDS AND PHRASES

During the 1940s and 1950s, the "BEEB" (as the British Broadcasting Corporation is affectionately known as in Britain), broadcast a radio show known as "The Brains Trust." During each of the 84 broadcasts, a panel of 4 erudite personalities attempted to answer questions that were put to them. Three of them anchored the programme and the fourth place was occupied by some well-known intellectual. One of the three respondents who anchored the programme began his response to any question by stating that "it all depends on what you mean by......." a particular word or phrase. He was C. E. M. Joad, a philosopher and psychologist who had written an excellent introduction to philosophy. His phrase became part of everyday language usage in the British Isles: even children would be heard using it. With a laugh of course!

Many years later during an interview for a senior academic post I used the same phrase in response to an interviewer who was asking me to comment on the philosophy of R. S. Peters. Professor Peters was responsible for making the study of educational philosophy something that had to be done in university departments responsible for the education of teachers. Among other things, he had said that "education was the initiation of worthwhile activity," or words to that effect [1]. Forty-five or more years have passed since then and it would be foolish to suggest that I can remember how the question was put, but I do know that I knew little or nothing of Peters work and that my response was to say "it all depends upon what you mean by education," etc. Of course I did not get the job!

Nowadays, I appreciate that the phrase merits some discussion. For example, it revolves around what you mean by "worthwhile activity." What is worthwhile activity?

I suspect that in any group of a dozen or so people, while some would give similar answers, others would give different answers as to what they perceived worthwhile activities to be. There would have to be some clarification, and the development of the ability to clarify is something that the activity of philosophy can encourage. But let us stay with the issue for a minute. Suppose we find some answers focus on worthwhile activities in the classroom while other responses refer to a range of activities from such things as gardening to going to a pop concert. We might argue that only the former are educational. But who are we to say that no learning takes place in the latter? So if we change the meaning of education to learning, and if we take it that more or less everything we have to do is worthwhile, by definition we arrive at something we know to be universally true, that is, that learning takes place all the time, contingent though it may be. Take another step and we begin to recognise that the system of formal education is a social artefact. Finally, we find that someone else is prepared to take all these arguments apart. That is what philosophers do. They take each other's arguments/systems apart—that is how philosophy moves on but never escapes from the arguments of the past.

Joad, or "Professor Joad" as he was known, was being a philosopher. He showed us that many statements require clarification if their meaning is to be understood.

Read the Wikipedia biography of Joad, or any other biography that you can call up, and you will find it was Joad who popularised philosophy, that he was a socialist, that he liked the ladies, that he liked mixing with the grandees, that he wrote prolifically, and that near his death bed he returned to Christianity. However, you will find little or nothing about his philosophical beliefs. Yet, among the long list of his publications you will find "*Critique of Logical Positivism*" [2]. This suggests that he may have allowed for metaphysics in his thinking which logical-positivism does not.

1.2   THE VIENNA CIRCLE

Logical positivism is the development of Bertrand Russell's [3] analytic philosophy by a group of persons who were primarily mathematicians and scientists working in Vienna. They became known as the Vienna Circle [4]. Their bible was "The Tractatus," published in 1921 by Ludwig Wittgenstein who had been a pupil of Russell's [5]. They were concerned with the "true" meaning of statements which it was argued could only be discovered by asking the question "what would he have to do to establish the truth or falsehood of this statement?" Because the formal methods they used for the verification of empirical questions were similar to those used in the sciences, they argued that only propositions in the natural sciences were true, and that it was impossible to say anything meaningful about ethics, aesthetics, and metaphysics. Thus, the clarification of meaning is confined to natural science. As such, it was in the tradition of British empiricism and positivism [6].

One of those who accompanied Joad as a regular panellist on the Brains Trust was the philosopher A. J. "Freddie" Ayer. He introduced the British public to "dogmatic logical-positivism" and the philosophical thinking of the Vienna Circle through his book *Language, Truth, and Logic* published in 1936 [7]. Ayer was then Professor of Philosophy at University College London. Like Joad, he had socialist leanings although he was not a pacifist. However, his logical positivism was quite clear. Only scientific statements can be proved to be either true or false and this implies a limitation on science, and therefore philosophy, since science has to be restricted to observable aspects of nature. Metaphysical and theological statements have no meaning, and the activities of philosophy become focused on how to replace ordinary language with more precise and standardised equivalents. This is the reason for mentioning it here, although as a philosophy logical positivism is no longer in the vogue that it was [8]. Nevertheless, in stating the case for a pluralistic approach to a philosophy of engineering, Carl Mitcham and Robert Mackey argue that a linguistic philosophy of engineering, as distinct from analytic philosophy, has a role to play in this pluralism [9]. In the context of engineering education, it is, as would be inferred in Section 1.1, concerned with the precise use of language in the design and implementation of the curriculum.

Although the average member of the public, and for the most part that is you and I, would not want to engage in the abstract conversations of philosophers on language, some things have trickled down into the public arena. We now appreciate that we have to be very careful in interpreting statistics that we find in newspapers. Nearer to home, we have become increasingly aware of the need to clarify meaning: we know that if the assessment questions we set in a public examination or the subsequent grading procedures are unclear there is the possibility that we will be taken to court. More pertinently, we know that if an instruction we give to a technician is misunderstood and leads to an accident that we are ultimately responsible for what happened. So we need to check that our instructions are understood and not misunderstood.

1.3   AMBIGUITY IN ENGINEERING EDUCATION

Since the year 2000, engineering educators in the U.S. have been required by the Accreditation Board for Engineering and Technology (ABET) to ensure that the programs they teach will achieve certain specified outcomes. They were revised, not without considerable debate, in 2016. Before they were first introduced in the year 2000, engineering educators were able to attend meetings that clarified the meaning of these outcomes. Two engineering educators, Yokomoto and Bostwick, argued that "secondary meanings of some words are sometimes used, such as using the term 'criteria' to describe the level of performance that students must achieve and 'outcomes' to describe the learning behaviours students must demonstrate" [10]. A more common definition of "outcome" is "result" or "consequence," and anyone attaching that meaning to the word will surely become confused in any discussion about writing measurable outcomes. Yokomoto and Bostwick said that the aims listed by ABET (Exhibit 1.1) were considered to be too broad to be assessed directly. In the tradition of *The Taxonomy of Educational Objectives* [11], they recommended that those aims should be broken down into smaller more measurable units. The essence of their argument was that accrediting agencies should explain the terms used, and use them consistently, and to this end they made a distinction between course outcomes and course instructional objectives. Again, such distinctions are debatable.

It is easy to fall into the trap of making ambiguous statements. For example, recently I wrote in a chapter of a book a modification of a statement that I had written in 1986 [12]. It was a definition of technology in which I now substituted "engineering" for "technology." The revised statement which related to the model is shown in Exhibit 1.2 reads, "Engineering is the art and science of making things that meet the needs of self and society. It is both an activity and a system that serves both individuals and society *that creates new problems for both*. Therefore, engineering literacy is necessarily interdisciplinary and a liberal study. *Engineering literacy is about the process of engineering whereas as technological literacy is about the products of engineering and their impact on society.*" It was the first phrase in italics that bothered the reviewer. He or she wrote: "*Meaning engineering creates new problems for both? Or the fact that it serves both individuals and*

Engineering programs must demonstrate that their graduates have:

(a) an ability to apply knowledge of mathematics, science, and engineering;

(b) an ability to design and conduct experiment, as well as to analyze and interpret data;

(c) an ability to design a system, component, or process to meet desired needs;

(d) an ability to function in multi-disciplinary teams;

(e) an ability to identify, formulate, and solve engineering problems;

(f) an understanding of professional and ethical responsibility;

(g) an ability to communicate effectively;

(h) the broad education necessary to understand the impact of engineering solutions in a global/social context;

(i) a recognition of the need for and an ability to engage in life-long learning;

(j) a knowledge of contemporary issues; and

(k) an ability to use the techniques, skills, and modern engineering tools necessary for engineering practice.

Exhibit 1.1: The list of program outcomes in Section II of Engineering Criteria 2000. The Accreditation Board for Engineering and Technology (ABET).

society creates problems?" [13] The ambiguity is immediately apparent. I was intending the former but, isn't the second point valid?

In the original statement the word technology was used because those of us who wanted to include some engineering science in the middle and high school curriculum in England failed to get our idea established. Western governments had begun to replace the industrial arts (woodwork and metalwork) with an approach that was based on design and make projects, and a syllabus based on a black box systems approach. Since then, as members of the Technological Literacy Division of the American Society for Engineering Education have pointed out, I have allowed the two literacies to become interchangeable when there are discernible differences. Some argued that I was confusing the issue so it became evident that there was a need for clarification and definition. At the same time, the Division also questioned the meaning of the term technology in its title in relation to its aims. A group led by John Krupczak gave separate and different definitions of engineering and technological literacy which has caused the division to change its name so as to embrace both engineering and technological literacy. This is why Krupczak et al. definitions are summarized in the second block of italics in the statement just considered [14].

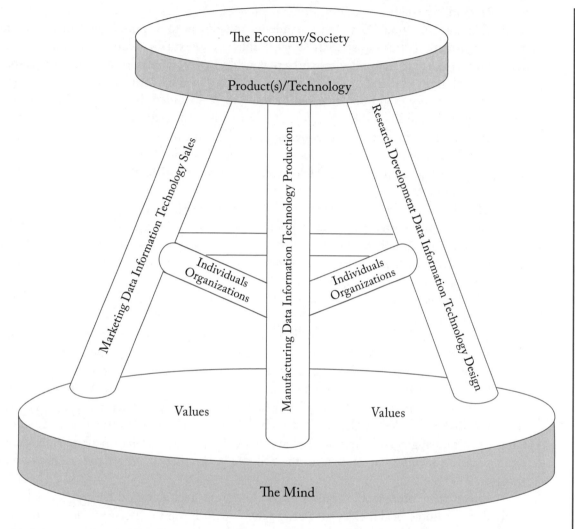

Exhibit 1.2: A model of the engineering processes engaged in the production of a technology (technological product).

1.4   DEFINING LITERACY AND TECHNOLOGICAL LITERACY

These definitions are helpful in so far as they clarify the differences between engineering and technology and demonstrate the breadth of curriculum issues that have to be considered. But the idea of literacy is also brought into contention since many would argue that it requires development. David Drew suggests just such a development. He writes, "Literacy means being able to understand information—'minimum competences.' And, yes, it is better to be literate than to be illiterate. But shouldn't we want students and others to acquire knowledge, not simply process information. This translates into being able to contextualize information. They must go beyond understanding how a device works to considering the implications of its use for society. Critical thinking is essential" [15]. But Drew thinks that while critical thinking and contextualizing technology might be perceived by some to be included in the term "literacy," a new term is required. Building on Krupzcak et al.'s distinction, he writes that we might differentiate consideration of the product ("technological literacy"), consideration of the process ("engineering literacy"), and contextual consideration of the implications ("technological knowledge" or perhaps "technological judgment"). He concludes that an interdisciplinary approach to the curriculum is required.

Cheville also defines literacy but in a different way [16]. He takes issue with me for he believes that while it is correct to frame technology as a necessary form of knowledge I "seemingly equate technological literacy with knowledge of technology specifically, and knowledge itself more generally," which while correct at a surface level "since the word 'literate' means having a knowledge of letters (i.e., literature)," limits the "definition of technological literacy by framing it as epistemic knowledge." He then argues this point by considering the etymological origin of the word technology. It originates in the Greek word "*techne*" which was one of Aristotle's five intellectual virtues along with *nous* (intellect), *sophia* (wisdom), *phronesis* practical judgment, and *episteme* (knowledge of things that are unchanging) [17]. *Techne* was fundamentally different to *episteme* since it was a virtue developed by the act of making, thus relating to artefacts that were made (came into being) rather than what existed naturally. Thus, from an Aristotelian perspective, the phrase "technological literacy" may be seen as either contradictory or as involving two related forms of knowledge. Despite today's belief in a divide between theory and practice, the latter definition aligns with Aristotle's views. I could have replied that my definition serves the latter since the opening statement links the two, *viz* "technology is the art and science of making things…"

Given that "proficient" means to make progress "technological literacy" should be expanded to "technological literacy and proficiency" [18]. From this perspective it is not enough to be literate in technology, rather those who wish to be proficient in technology must also develop the skill to create, for as Aristotle says *techne* is concerned with those things "whose origin is in the maker and not in the thing made" [19]. Cheville argues that this line of thinking eliminates the distinction between engineering and technological literacy.

It might be argued that this is not achievable with the broad audience that is the focus of a programme in technological literacy. Cheville, while acknowledging that this is a difficult issue, says that "while education can provide contextual literacy, proficiency requires not only the ability to choose the right purpose, but to be able to act in the right way. It may be argued that such skill must be gained through experience so that effectively teaching both technological literacy and proficiency requires students to be exposed to a range of contexts and problems" [20].

Wrestling with meaning is important and in the context of this example throws up issues about what it means to be technologically literate in terms of knowing and doing that are far from resolved. It is a matter that is further complicated by the different languages that have to be spoken.

1.5   THE LANGUAGES OF ENGINEERING

The definition of engineering/technology used to establish the previous discussion was originally employed in collaboration with the model shown in Exhibit 1.2. It was intended to show the interrelationships between the areas of engineering and the resulting achievement of a technological artefact for society and the economy. It also shows that its comprehension requires understandings of different languages, including our own and languages that are foreign to us. For example, both engineering and technology have to function within legal constraints legal and otherwise, imposed by society and the environment. Speaking to those is to speak a different language to engineering.

The base of the model represents the person. The mind that supports the whole activity is the source of our values, beliefs, and technical understanding; it is the source of our attitudes and opinions to the different social systems in which we find ourselves, it is the driver of our actions. That is how this dimension of the model has been presented on several occasions but it is also the source of our ideas and creativity. Understanding how our beliefs and values (moral and otherwise) are formed is important to our conduct as engineers and individuals but it belongs primarily to the domains of philosophy and theology which are different languages.

The three legs of the stool represent the technological aspects of engineering: research and development, data acquisition, information technology; design; manufacturing data and production; and marketing data and sales. The first two legs are the domains of engineering science, design, and manufacturing. The third leg is the knowledge domain of business, legal, and economic understanding. Supporting the legs are the trusses that represent individuals and the way the organisation is structured. These are the domains of organisational behaviour and behaviour in organisations. The seat represents the economy and society within which the product is placed. Presently, as we all know, there is much concern with the impact of IT on society and the individual, especially children, and whether or not it will come to control us [21].

It is quite evident that an engineer or a manager, indeed each participant, producer and user, has to learn several languages if they are to understand the "meanings" that each person brings to the activity of engineering.

Bucciarelli, who evaluated designers at work, explained that the participants worked within different *object worlds* each of which had a language proper to that world. "The languages of different object worlds are different; their proper languages are different. In another world apart from the structural engineer, the electronics engineer speaks not of stress and strain bit of power, voltages and currents, analogue and digital, resistance and capacitance. The mathematics may appear similar—there are strict analogies that apply in some instances—but the world of electronics is different, populated by different variables, time scales, units, scientific law and principles of operation. So too different kinds of heuristics, metaphor, norms and knowledge codified, tacit and know-how" [22, p. 16].

Follow Bucciarelli for one more paragraph and the challenge of technological literacy for the public becomes clear. He writes, "Participants within object worlds function as *élites*. But the case is different from that pictured by the philosopher Hilary Putnam. Object worlds divide the design task into different, but not independent, kinds of effort so one can say that there is a 'division of linguistic labour' but the distinction is not that there is one group, an *élite*, that know the full meaning, has a God's eye view of the object of design and another group with but a less sophisticated, common understanding of the design task. Rather there are multiple *élites*, each with its own proper language. It is in this sense that different participants within different object worlds with different competencies, responsibilities and interests speak different languages. Crudely put, one speaks structures, another electronics, another manufacturing processes, still another marketing, etc." [23, p. 19].

Given that every member of the public has some degree of involvement in the Boeing 737 Max crash, to the extent that we may have to decide whether or not to fly on it when flights are resumed, the question becomes how can technological literacy help us, if at all, make that decision? Bucciarelli's understanding of design would lead us to suspect that there is no one who has a command of the whole of the detail of the aircraft. So we have to look to other object worlds for a solution to the problem. These are to be found among those who made the decision to proceed with the design in the first place. There are some quite simple questions that should be asked about the overall design that was approved [24]. In the UK, similar questions apply to the Grenfell fire tragedy [25].

1.6   AN ALTERNATIVE SCENARIO: THE THEATRICAL

An alternative but complementary scenario to Bucciarelli's is to consider the scene (particular social system) as theatre. Each of the participants is a role player who speaks a different, or a number of different language(s). These languages relate to the roles they have to play. Each role player occupies a plurality of social systems and much of life is spent trying to create an equilibrium between these systems [26]. For example, the work system has to be balanced with the career system and both with the family system. Role players are in exchange with the environment and they try to shape or adapt the system in which they find themselves. In some circumstances they will find themselves in conflict with other role players, and at other times because of ambiguity, they may find themselves in mental conflict with themselves. Circumstances and the person will dictate the individual response to conflict and tension. By themselves they are neutral but sometimes tension will be beneficial to the individual and at other times harmful [27]. Clearly, people need to be able to adapt and shape their environment in the systems they occupy [28].

While much attention is now paid to teamwork it is likely to be the role players with which a given role player works that are impacted by their relationships. How the individual role players exercise power over each other is of the utmost importance to the effectiveness of the organisation [29]. From the perspective of the curriculum it is important that students understand the roles and the systems in which they function, and that their work in teams should be analysed from this perspective, for "role systems exist for interpersonal relations, which depend on influence, which depends on power. Power is a resource, while its effect is influence. The effects of influence are cooperation, conflict, love, hate, fear, jealousy, and all those other emotions encountered in role systems" [30]. The emotion that is of paramount importance in learning, be it in college or at work is "motivation."

Macmurray is surely correct when he argues that "emotional reflection is the primary mode of reflective activity" [31, p. 194]. Macmurray suggests three modes of reflection in which the agent's, or as I interpret or develop this thinking, the role player's reflections lead to action. The first is the *pragmatic mode* which to all intent and purpose is what we commonly call scientific. In this mode the role player tries to generalise the results in order to provide for more effective action in the future. Left to its own it is positive. But role play cannot be effected without emotion. Macmurray calls the emotional significance of an action the *contemplative mode*…. "We may say of the emotional mode of reflection that it seeks to determine the world as an end itself, or rather as a manifold of ends. As we called intellectual knowledge of the World-as-means, so we may describe emotional knowledge as the knowledge of the World-as-end" [32, p. 194]. The personal mode reflects on whether the action is right or wrong and, is thus, concerned with moral behaviour. Thus, all our actions are judged by us in terms of their pragmatic, emotional, and moral contents, and it against these that we may choose to select our curriculum objectives (see Section 2.6 on screening aims). They represent in everyday parlance and attempt to describe what many people would understand by the term "the whole person" which they would hold is a major aim of education particularly as expressed in Newman's *Idea of a University* [33].

1.7   TO SUMMARIZE

Joad's rhetoric was not idle. I make no apology for greatly simplifying the philosophical debate about logical positivism and more generally the analytic tradition in Britain because it caused the public to understand that much care should be taken to ensure that the "meanings" they wished to convey are understood in the way they wished them to be understood [34]. Wittgenstein did not consider there was any such thing as "pure" thought. It is the language we possess that enables us to think. It is a way of life. Bucciarelli explained the design process as being a "division of linguistic labour." All language is shared but that is a consideration for another day.

This short journey into the meaning of meaning was inspired by a popular British philosopher's persistence in a series of radio broadcasts to query the meaning of words and statements. It led us to the many object worlds and many languages of those involved in the design process and, when related to technological literacy these include the customer. More generally, Joseph Pitt reinforces what Bucciarelli shows, namely, that "philosophy is not all about how language works, but how people use it to communicate better and solve problems affecting their immediate lives. It is not all about logic, but also about how people reason informally and why that is, in the long run, more important than producing a consistency theorem. It is not just about the structural flaws in various ethical theories, but about what kind of a world we should be building in order truly to live the good life. In short, is about people. It is about how all of us relate to each other and to the world. And if it is about all of us, then it necessarily includes engineers" [35]. All points that have been developed by Alan Cheville [36].

An alternative and complementary approach to Bucciarelli's based on role playing and systems was presented. It led to a possible description of that aim of education that is to educate the whole person.

This chapter illustrated the importance of clear meanings in the design of assessment and learning, thus in curriculum design. Such discussions can have an impact on what we think the aims of the curriculum, indeed education should be. Unfortunately, as the next chapters show, it is very difficult to get educators to agree on the terms that should be used.

1.8   POSTSCRIPT

In Section 1.6, the work of two British academics was introduced. They will be referenced throughout this text. The first mentioned was John Macmurray who was professor of philosophy at the University of Edinburgh having held a similar position at University College London. He was the first person to give popular lectures on philosophy for the BBC. The second academic mentioned was Saint John Henry Newman who, having set out his idea of a university in lectures and writing, founded the Catholic University of Ireland which eventually became University College Dublin. Both were Christians but of very different persuasions: Macmurray, a Quaker; Newman, a convert Catholic. In that tradition, both were concerned with the "personal."

In Section 1.6 their thinking was used in support of the argument that individual's in adapting to and shaping their environment, emotions, or the domain of the affective is of equal importance to the domain of the cognitive in personal development. Working and managing in teams, as engineers are often asked to do, requires the development of good relationships, and through such relationships, Macmurray argued, we come to know who we are. Newman also attached considerable importance to the relationships established in college as one of the mechanisms of development, and a recent study at Hamilton College in the U.S. suggests relationships are the single most important factor in success (Section 3.7).

More generally, Macmurray might well be considered to be the engineer's philosopher for his was a theory of doing, not a theory of thinking. Thus, the reflective practice referred to in Section 1.6 is necessarily related to our practical life, for all our "theoretical activities have their origins, at least, in [our] practical requirements" [37, p. 21]. Newman's reflective activities certainly had their origin in the need to solve practical problems but the concern here is with his view of the universality of knowledge (end note—In Newman's time "universal" did not mean "all inclusive" as it does today but "turned into one" being derived from "*uni-versum*") which forces a person to look at "the whole body of things, and therefore the true character of a university is not that it teaches all the sciences but that whatever it teaches is taught in a spirit of universality" [38]. The diagram in Exhibit 1.2 shows the body of things to which a technologically literate person has to give thought. In one way or another the skill involved in technological literacy is recombining subjects so that the whole is apparent. Culler gave an example related to Westminster Abbey which better illustrates the point if Notre Dame de Paris is substituted for the abbey, because those ultimately responsible for restoring the cathedral after a fire in 2019 nearly destroyed the building, have to know it thoroughly as a material object. "If we wished to know a single material object for example, [Notre Dame de Paris]—to know it thoroughly, we should have to make it the focus of universal science. For the science of architecture would speak only of its artistic form, engineering of its stresses and strains, geology of its stones, chemistry and physics of the ultimate constitution of its matter, history of its past and literature of the meaning which it had for the culture of the people. What each one of these sciences would say would be perfectly true to its own idea, but it would not give us a true picture of [Notre Dame de Paris]" [39]. Newman argued that to get this true picture the sciences would have to be recombined and this recombination is the object of university education. If this is achieved, technological literacy achieves this goal. A recent case study of the Boeing 737 Max accidents attempts to demonstrate this point [40].

1.9   NOTES AND REFERENCES

[1] Peters, R. S. (1964). *Education as Initiation*. London, Evans. 1

[2] Joad, C. E. M. (1960). *Critique of Logical Posivitism*. London, Gollancz. 2

[3] For a brief introduction to Bertrand Russell and analytic philosophy, see Magee, B. (1998). *The Story of Philosophy*, London, Dorling Kindersley, pages 196–207, which also includes a substantial section on Ludwig Wittengenstein. See also Ref. [8]. 2

[4] Uebel, T. (1995). Vienna Circle in R. Audi (Ed.). *The Cambridge Dictionary of Philosophy*. Cambridge, UK, Cambridge University Press. 2

[5] Wittgenstein, L. (1922). *Tractatus-Logico-Philosophicus*. Transactions on Ogden, G. K., London. Routledge and Kegan Paul. This was the only book published in his lifetime. His other major work was *Philosophical Investigations*, 1953. Trans. G. E. M. Anscombe. Oxford. Blackwell. 2

[6] (i) Empiricism. Experience is the only source of knowledge. 2

(ii) Positivism. Described by Alan Lacey as being "a movement akin to empiricism and naturalism introduced by Comte a French sociologist (to use a term he himself invented)," (p. 742, 2nd ed., *Oxford Companion to Philosophy*). This philosophy proclaims that all knowledge is based on sense-experience, and the fact that all learning is based on the description and explanation of empirical facts. Thus, there are no differences between the sciences and the social sciences in the methods they use. Its precursor is empiricism.

(iii) For a summary of the various—isms in philosophy see Grimson, W. (2014). Engineering and philosophy in *Philosophical Perspectives on Engineering and Technological Literacy*. Handbook no. 1 Technological and Engineering Literacy/Philosophy Division of the American Society for Engineering Education, Washington DC.

[7] Ayer, A. J. (1936, reptd 2001). *Language, Truth, and Logic*. Harmondsworth, Penguin. 2

[8] Fotion, N. G. (2005). Logical positivism in Honderich, T. (Ed.). *Oxford Companion to Philosophy*. Oxford, Oxford University Press. 2, 11

[9] Mitcham, C. and Mackey, R. (2010). Comparing approaches to the philosophy of engineering: including the linguistic philosophical approach in I. van de Poel and D. E. Goldberg (Eds.). *Philosophy and Engineering. An Emerging Agenda*. New York, Springer. 2

See also McCarthy, N. in the same volume *The World of Things not Facts*, pages 265–273 which is an essay on the relevance of Wittgenstein to a philosophy of engineering.

It should be noted that Mitcham and Mackey provide useful introductory descriptions to six currents in contemporary philosophy. These are: (1) phenomenological, (2) post-modern, (3) analytic philosophy, (4) linguistic philosophy, (5) pragmatism, and (6) thomism. They make a sharp distinction between analytic and linguistic philosophy which is not evident in the above text, a viewpoint that seems to be supported by the distinguished American philosopher Richard Rorty.

[10] Yokomoto, C. F. and Bostwick, W. D. (1999). Modeling the process of writing measurable outcomes for EC 2000. *Proc. Frontiers in Education Conference*, 2(11b1):18–22. Piscataway, NJ, IEEE. 3, 29

[11] Bloom, B. (Ed.) (1956). *The Taxonomy of Educational Objectives. Handbook 1 Cognitive Domain*. New York, David McKay. 3

The taxonomy is one of the most widely referenced educational texts of all time. Many engineering educators have used it, and it is claimed to have had a significant influence on education world-wide. It provides a hierarchical framework for categorizing educational objectives for use in test and curriculum design. The hierarchy is made up of six domains and in the text these are accompanied by sub-categories. These are expressed in terms of the behaviours a student would demonstrate if they possessed the skill required. Thus, a student has the "ability to produce a unique communication" is one of the categories of the domain of synthesis. The six domains are Knowledge, Comprehension, Application, Analysis, Synthesis, and Evaluation. It was open to much criticism (see Chapter 2 of

Heywood, J. (2005). *Engineering Education. Research and Development in Curriculum and Instruction.* Hoboken, NJ, Wiley/IEEE).

The authors of the *Taxonomy* made it clear that their descriptors were of the "outcomes" of education. The influence of this approach can be seen in the ABET 2000 EC list of outcomes, and the outcomes for the different levels of the Bologna agreement. But those authorities did not use the *Taxonomy* and they preferred, as have very many other educational authorities, to use the term "outcome" rather than objective.

The taxonomy was revised in 2001. The Knowledge Domain was sub-divided into four dimensions—Factual knowledge, Conceptual knowledge, Procedural knowledge, and Meta-cognitive knowledge. Six cognitive process dimensions were included—Remember, Understand, Apply, Analyze, Evaluate, and Create.

Anderson, L. W. et al. (Eds.) (2001). *A Taxonomy for Learning Teaching and Assessing. A Revision of Bloom's Taxonomy of Educational Objectives.* New York, Addison Wesley Longman.

[12] Heywood, J. (1986). Toward technological literacy in Ireland: An opportunity for an inclusive approach in Heywood, J. and Matthews, P. (Eds.). *Technology, Society and the School Curriculum: Practice and Theory in Europe.* Manchester, Roundthorn. 3

[13] Personal communication, June 2013. 4

[14] Krupzcak, J. et al. (2012). Defining engineering and technological Literacy. *Proc. Annual Conference of the American Society for Engineering Education.* 4

[15] Drew, D. E. (2017). Moving the needle from literacy to knowledge in Cheville, A. (Ed.). *Philosophical Perspectives on Engineering and Technological Literacy IV.* A publication of the TELPhE Division of the American Society for Engineering Education. Washington, DC, ASEE. 6, 13

[16] Cheville, A. (2017). Technological literacy without proficiency is not possible in Cheville, A. (Ed.). *Philosophical Perspectives on Engineering and Technological Literacy IV.* A Publication of the TELPhE Division of the American Society for Engineering Education. Washington, DC, ASEE. 6, 13

[17] Aristotle, *Nicomachean Ethics*, VI-4. 6

[18] *Loc. cit.* Ref. [15]. 6

[19] *Loc. cit.* Ref. [16]. 6

[20] *Loc. cit.* Ref. [15]. 6

[21] *The Times* of June 22, 2019 carries two articles: (i) Macintyre, B. Will the march of the machines lead to our enslavement? p. 16–17. (ii) Turner, J. Porn is warping the minds of a generation. p. 15. 7, 15

[22] Bucciarelli, L. L. (2003). *Engineering Philosophy*. Delft. DUP Satellite. 8

[23] *Ibid.* 8

[24] Heywood, J. (2021). The concepts of technological literacy and technological competence examined through the lens of a case study concerning the Boeing 737 Max accidents. *Proc. of the Annual Conference American Society for Engineering Education*. Paper 33315. 8, 16

[25] Heywood, J. and Lyons, M. (2018). Technological literacy, engineering literacy, engineers, public officials and the public. *Proc. Annual Conference of the American Society for Engineering Education*. Salt Lake City. 8

[26] Burns, T. (1966). On the plurality of social systems in Lawrence, J. R. (Ed.). *Operational Research and the Social Sciences*. London, Tavistock. 8

[27] Heywood, J. (1989). *Learning, Adaptability and Change. The Challenge for Education and Industry*, Chapter 3, London, Paul Chapman/Sage. 8

[28] The ability to adapt to and shape the environment is Sternberg's definition of intelligence. See Sternberg, R. J. (1984). *Beyond IQ. A Triarchic View of Intelligence*. New York, Cambridge University Press. See also Heywood, J. (2018) *Empowering Professional Teaching in Engineering*. Journey 15, San Rafael, CA, Morgan & Claypool. 8

[29] Youngman, M. B., Oxtoby, R., Monk, J. D., and Heywood, J. (1978). *Analysing Jobs*. Aldershot. Gower Press write; "All persons direct or control and thus use managerial skills at one time or another. Thus, in the firm concerned in our study everyone was involved in direction and control and, hence, in management. Job satisfaction is to some degree a measure of the extent to which individual needs for direction and control are satisfied. This in turn as […] shows is as much a function of personality as it is of personal history, ability and interest. What is an acceptable goal to one person will not be to another; some want to be stretched, others want a strict routine. No two persons in a section will be exactly alike. In the same section there will be both aggressive people and timid people who, if they are taken outside the sphere they are capable of controlling, will have to be supported. Sometimes people of high ability will find themselves in this situation." 9

"A person is a psycho social system. Within the boundaries of that system, most individuals wish to be 'organic' to use a term first suggested by Burns and Stalker*. They wish to be able to take actions and decisions as well as mature. The boundaries of these psycho-social systems arise as a function of the needs of the job and the needs of the person. When these are matched for each person in the organisation a hierarchic system becomes structured by individuals who are organic within their own system. The system itself becomes organic if it can respond to the needs of individuals. Both systems have to be self-adjusting."

*Burns, T. and Stalker, G. M. (1961). *The Management of Innovation*. London, Tavistock.

[30] Hunt, J. W. (1979). *Management. People at Work*. Maidenhead, UK, McGraw Hill cited in Ref. [21]. 9

[31] Macmurray, J. (1957). *The Self as Agent*. London, Faber and Faber. 9, 15

[32] *Ibid*. 9

[33] Newman, J. H. (1852, 1923 impression). *The Idea of a University. Defined and Illustrated*. London, Longmans Green. 9

[34] Magee, B. (2001). Writes in a section of his book called "*Common Sense*," that "meanwhile (in the 1930s) in Britain a near-contemporary and lifelong friend of [Bertrand] Russell's called G. Moore had been pursuing the analysis of statements in ordinary language using neither science or technical logic as his yardstick but common sense [...] into a mode of philosophy that was eventually to displace Logical Positivism. It became known as 'linguistic philosophy' or 'linguistic analysis,' and its criterion was the ordinary use of language. The Logical Positivists had been mistaken said the linguistic analysts in trying to force the straight jacket of scientific standards on all forms of utterance. Umpteen different sorts of spontaneous discourse go to make up human life, and each one has its own logic. Philosophical problems arise when a form of utterance appropriate to one mode of discourse is mistakenly used in the wrong context" (pp. 200–201). Magee writes of Wittgenstein that in his later philosophy "linguistic analysis achieved its ultimate degree of refinement" (p. 202). 9

Magee, B. (2001). *The Story of Philosophy*. Dorling Kindersley, London.

[35] Pitt, J. C. (2013). Fitting engineering into philosophy on Michelfelder, D. et al. (Eds.). *Philosophy and Engineering: Reflections on Practice, Principles, and Process*. New York, Springer. 10

[36] Cheville, R. A. (2019). *Becoming a Human Engineer. A Critique of Engineering Education*. Cambridge UK. 10

[37] *Loc. cit*. Ref. [31]. 10

[38] Culler, A. D. (1955). *The Imperial Intellect. A Study of Newman's Educational Ideal*, pages 180–182, New Haven, CT, Yale University Press. 11

[39] *Ibid.* 11

[40] *Loc. cit.* Ref. [24]. 11

CHAPTER 2

The Curriculum

2.1   TERMINOLOGY IN EDUCATION

There is no agreed terminology in education that stands the test of time. The prime example of this is the use of the term "*objective*" which was used in the most referenced education book of all time *The Taxonomy of Educational Objectives* [1]. In spite of the fact that the authors of the first handbook on the cognitive domain are quite clear that their statements of behavioural objectives that say what a student will be able to do are statements of learning outcomes, present day usage prefers the term "*outcomes*." The authors of the revised taxonomy published in 2001 dropped the use of the term "*objective*." The title of the revision is *A Taxonomy for Learning, Teaching, and Assessing*, with the sub-title *A Revision of Bloom's Taxonomy of Educational Objectives* [2]. However, the authors remain clear that objectives "describe ends-intended results, intended outcomes, intended changes" [2, p. 17]. They say "that today's terminology is driven by the current emphasis on school improvement through standard-based education (in the U.S.). At the heart of the standards-based movement is the state-level specification of intended student learning outcomes in different subject matters at each grade level" [2, p. 18]. They note that the standards fit nicely into the three levels of objectives described: global, educational, and instructional.

I have no idea how the term outcome came to replace objectives in higher education. The first references I have to its use in education are in two documents prepared for the British Employment Department (equivalent U.S. Department of Labour) by Sue Otter in 1991 [3] and 1992 [4]. In the first report she said that objectives were used by course designers whereas specifications of learning outcomes were made by teachers. In parallel with Otter's publication was a book by Gilbert Jessup that described the outcomes system that was being introduced in vocational education in the UK [5]. Otter included a taxonomy of outcomes for engineering education by R. G. Carter that had been published some years earlier [6, 7]. His taxonomy is of particular interest because it is about the personal. That is, "the personal qualities to be possessed by students than the learning experiences by which they are to be developed." It focuses, in particular, on how the objectives are to be learnt and structured for their development. His categories depended as much on the affective domain as they did on the cognitive. They were: Mental skills, Information skills, Action skills, Mental quality, Attitudes and values, Personal characteristics, Spiritual knowledge, Factual knowledge, and Experiential knowledge.

It is clear that some of these outcomes could not be achieved without what came to be called *active learning*. At that time they promoted engineering educators with a profound challenge, and they marked a major change in how the curriculum should be considered.

2.2   FROM TERMINOLOGY THROUGH IDENTITY TO AN OPERATIONAL PHILOSOPHY

The reasons for terminological disagreement in the social sciences are many, one of them undoubtedly is the search for identity. There is much kudos to be gained for a scholar who comes up with a term that becomes vogue. I have always missed out on such matters. Clearly, in a paper that I published in 1961, I was writing about discovery learning, but my explanation was cumbersome and not accompanied by a descriptive single term. I could equally well have coined the phrase "problem-based learning." Discovery learning was popularised by Jerome Bruner in a paper in the *Harvard Educational Review* in the same year (1961) [8]. The term seems to have originated with John Dewey but as Mani Mina and others have shown [9] Dewey's model is now called inquiry learning, and the term discovery has been lost.

The terms we use contribute to our identity, and when we seek out groups we adopt the terms that describe the identity of those groups. Currently, we read of identity crises and identity politics here, there and everywhere. No profession has been more infected by identity problems than the engineering profession.

For example, Priyan Dias has distinguished between three crises related to the engineers influence, role, and knowledge. He illustrated them by three questions, each of which related to a domain philosophy, and in turn relate to the curriculum. They are, "Are engineers doing more harm than good?", "Are engineer's scientists or managers?", and "Is engineering knowledge theoretical or practical?" Dias suggested that answers to the first question belong to the domain of ethics, the second to the domain of ontology, and the third question to the domain of epistemology. He argued that these questions "are valid for engineers in most if not all societies. It is the duality posed in the questions that creates the angst. It is the undermining of self-worth or social value inherent in the questions that constitutes the crisis" [10].

Answers to these questions certainly have a bearing on the design of the curriculum. For example, in the late 1960s, the training regulations of The Council of Engineering Institutions and the Engineering Industries Training Board sharply distinguished between the roles of the engineer and manager, to the detriment of the former. At the same time, there was an acknowledgement that some introduction to the area of management in undergraduate courses would be useful. In the Council of Engineering Institutions examination this was recognised by a compulsory examination paper in "The Engineer and Society." In university programs this was sometimes achieved by the addition of course; it was not seen as something that could or should be integrated in the problem as could be achieved by problem-based learning. As for the third question, universities took the view that engineering was essentially theoretical. Dias wrote that "most university programs in engineering are filled with theoretical subjects that are largely mathematics in disguise."

The operational or working philosophies that sustain university programmes derive from the values and beliefs of teachers and taught, and such is the motivation that they sustain, it is very difficult to change the curriculum without a more fundamental debate that challenges these assumptions [11].

The fundamental questions that Dias's conflicts of identity raise are "what is engineering?", and, therefore, "what is an engineer?" Answers to both depend on what engineers actually do, and this may be determined by philosophical analysis or job analysis, or both.

The view that engineering is applied science created difficulties for engineering educators and on two occasions they were forced to publish statements that distinguished engineering from science (physics). Subsequently, in 2004 Samuel Goldman pointed out that the philosophy of science and the philosophy of engineering were two different things [12].

He wrote "Mathematics is paradigmatic of what has been most admired in Western 'high' culture, namely reasoning that is abstract, necessary, and value free; and problem solutions are universal, certain, unique, and timeless. Historically, 'demonstration' meaning mathematico-deductive argument is the form of reasoning that the most respected Western philosophers from Plato and Aristotle to the early Wittengenstein have striven for, rejecting reasoning based on the probable, the concrete and the contingent." And science is based "the concept of 'necessity' which is cognate with concepts of 'certainty,' 'universality,' 'abstractness,' and 'theory.'"

Academic engineering in its search for intellectual status has firmly rooted itself in this tradition. Real engineering (the engineering done in industry) "is by contrast characterised by wilfulness, particularity, probability, concreteness, and practice." Manufacturing industry, and therefore, its training is rooted in the contingent. "Engineering is contingent, constrained by dictated value judgements and highly particular. Its problem solutions are context sensitive, pluralistic, subject to uncertainty, subject to change over time and action directed" [13]. It is rooted in a tradition of philosophy that was prior to the Socratic philosophers and continued through to the American pragmatists, in particular, John Dewey. It represents a different way of thinking.

This view may explain, to some extent, why young engineering graduates have difficulty in adapting to interest, and why some engineering students in cooperative (sandwich) courses find difficulty in adjusting to periods in industry, or vice versa. It represents a different way of thinking. Expressed in this way the student engineer is being asked to learn to think and act in two quite different ways. They are being asked to deal with two different dimensions of reality which is a tall order, and for which some induction into concepts of knowledge and learning that is followed up throughout the course would seem to be necessary.

One difference that student engineers or new graduates find is that much use is made of heuristics which is at the heart of the engineering method described by Billy Vaughan Koen [14]. According to Koen, a heuristic is anything that provides a plausible aid or direction in the solution of a problem but is in the final analysis unjustified, incapable of justification, and potentially fallible. Common heuristics used in engineering are (1) "Those that aid in the allocation of resources such as the admonition *to allocate resources to the weak link.*" (2) "Others that control the amount of risk such as the admonition *always give yourself a chance to retreat.*" (3) "Still others that indicate the appropriate attitude for an engineer to take when an encountering a problem such as the admonition to *Work at the margin of solvable problems*" [15].

Some affirmation of these views of engineering are to be found in a task analysis of the work done by engineers reported in 1978 [16]. That study did not find that there was a unique skill (ability) that might be called management, thus calling into question the separation of engineering from management in the training regulations of the Council of Engineering Institutions and the Engineering Industries Training Board [17]. Much more recent work by Trevelyan also supports this understanding [18].

To be fair, some preparation for the experiential learning that a student faces in industry can be provided by project- and problem-based learning, which is one reason why such techniques receive strong support in medical education.

It is with the curriculum as an instrument for reconciling the identity conflicts that engineers have that the next section is concerned. That requires somewhat more than an operational philosophy, as will be shown in the following sections.

2.3 DEFINING THE CURRICULUM: THE CURRICULUM AS PROCESS

As has been demonstrated, the terms used may be ambiguous and tied to local and/or national custom. That seems to be the case with the term "curriculum." It is used as a function of the context in which it is employed. At one level it is used to describe all subjects (contents) that comprise the course of studies that an individual pursues (e.g., civil engineering, computer engineering, and electrical engineering). At a second level, it is sometimes used in respect of a particular area of study (e.g., aerodynamics, electromagnetic theory). In both senses it may embrace teaching and assessment. In higher education these areas of study are sometimes called a subject or a course. Often, the curriculum refers to a list of content (topics to be covered). Such lists are sometimes called "The Syllabus." A syllabus is not the same thing as a two or three line description of a course that might appear in a university catalogue.

In this chapter, "curriculum is defined as the formal mechanism through which intended aims are achieved" [19]. Since educational aims are achieved through learning, the curriculum process is described by those factors that bring about learning. Thus, both learning and instruction are similar processes to the curriculum process. Because it is a complex process it is not easy to demonstrate visually. There is no agreed model. Sometimes when I have produced my model the person with whom I am conversing draws their own model. Indeed, my first diagram which has gone through several iterations [20] was triggered by diagrams due to E. J. Furst (1958) and J. F. Kerr (1963) [21, 22]. The iteration I like best is shown in Exhibit 2.1. It is in circu-

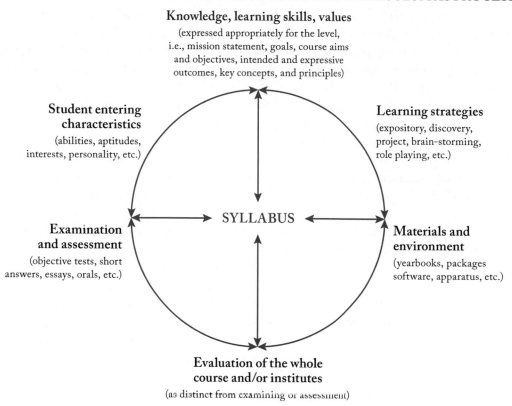

Exhibit 2.1: A model of the assesment-cirriculum-learning-teaching process (Heywood, J. and M. Murray, 2005). Curriculum-Led Staff Development. Towards Curriculum and Instructional Leadership in Ireland. *Bulletins of the European Forum on Educational Administration* No. 4. (Sheffield Hallam University. School of Education for EFEA).

lar form because the curriculum process is not linear. The late Georgine Loacker Chairperson of the Assessment Council at Alverno College did not like the model because, in her view, it did not display the dynamic nature of the curriculum. My revision to meet her complaint [23] never quite satisfied me because the diagram had to be accompanied by a lengthy explanation. It should be evident that a similar model can be used to describe the planning of a lecture (class) or series of lectures (classes).

Another criticism is that the model does not really show the impact of accreditation and quality assurance (QA) procedures on teaching and learning and higher education. The recent controversy over ABET's proposals to change their criteria shows just how important these procedures are. Arnold Pears and his colleagues [24] argue that QA can impact on the process in four areas that are a function of how the aims of education are perceived. These are: "1. *Service*—where identifying who is served, the customer is crucial. *2. Process*—where the focus is on formative structures in education. 3. *Body of Knowledge*—where the focus is on the knowledge components in education. 4. *Holistic personal; development*—where there is a focus on professional identity." They appreciate that these are overlapping fields and that each depends on what the purpose of education is. This reinforces the assumption made in this text which is, that curriculum design begins with a clear understanding of the purposes of the higher education are. That discussion would necessarily consider these four perspectives. Pears and his colleagues made the point that while broader definitions of quality are required they must be supported by an embedded culture of quality in the institution. Maura Borrego and her colleagues made this same point with respect to assessment [25]. These two dimensions of the curriculum process are, of course, closely related.

One way of modifying the model would be to insert an outer ring with inward pointing arrows, and labelled external institutions.

What the models do show is the dependency of the dimensions of the process on each other. This was little understood until, in the UK at least, research on assessment came to be taken seriously. An early axiom was that students would do what was required of them to achieve good marks in their examinations and assessment. Swedish research by Marton and Säljo showed that the style of assessment could cause either deep learning or surface learning. These concepts are self-explanatory [26]. Clearly, universities wish to encourage deep learning.

Prior to their 1976 report it had become clear, contrary to the position held in England by the Matriculation Boards [27], that examinations impacted on teaching. The view taken by the Boards was that their examinations should test the syllabus in the subject examined, and not interfere with the teaching. The fact of the matter is that they did, and do. Teachers predicted what questions would be asked in the examinations, and appropriately directed their teaching to prepare their students for those or similar questions. However, while it seemed clear that the objectives to be tested would dictate not only what was taught but the time available for its teaching, it is by no means clear that this was fully understood.

Such views impacted on D. T. Kelly and this writer when we were charged with designing the coursework for an examination in Engineering Science at "A" (Advanced) level, a public examination for entrance to university [28]. The examination comprised six hours of written exercises spread over 2 three-hour periods, and during the two-year course, the assessment of coursework that comprised experimental investigations and a project. The general idea was that the combination of coursework and the written exercises would assess skills used by engineers in their work. The model was very much that of the engineer as applied scientist partly because the subject had to be the equivalent of physics for its results to be used in the university selection procedure. A major difference between this approach and examinations in other science subjects was the inclusion of the project and experimental investigations. New approaches to the design of the questions and style of the written examinations. The assessments for the coursework used semi-criterion referenced scaled rubrics. In the case of the project the skill domains assessed were planning, execution, design, use of resources, and critical review. A final question sought to elicit from the first assessor (the teacher) how much the candidate had exercised initiative and judgement in pursuit of the project. Students were not allowed to begin their project until they had satisfied the examiners that they were competent to pursue the topic they had chosen. To achieve that goal students had to submit a project outline which included an analysis of the problem, the practical problems to be solved, possible solutions, resources, and a time table.

We argued that it was also desirable to test the skills of project planning in a written paper in which the students would be asked to carry out the initial phase of a design project. We expected that there would be a high correlation between performance in coursework and the written paper on project planning. In our first published paper, we reported that in each of the three years that the examination had been set, the correlations between these two dimensions were the lowest of all those obtained between the sub-dimensions of the assessment. We put this down to the fact that there was no formal syllabus requiring the teaching of engineering design, and that in consequence students were required to learn the associated skills by osmosis. At the time in the UK (the beginning of the 1970s), the idea that engineering design could be taught was contentious [29].

Subsequently, in Ireland we were able to more precisely pin point this difficulty in a project in which teachers designed an examination to test mathematical problem solving, but did not teach their students how to answer such questions [30]. Performance was poor. In another study, we affirmed that when students were shown examples of what was required performance improved. In that case, we demonstrated how there was a considerable improvement in performance when student teachers were shown how to use action research in their classrooms to solve teaching problems they were meeting [31], when compared with graduate student teachers solving similar problems without any prior examples.

In sum, instruction, the resources used and assessments have to be designed to meet the objective(s). In the U.S. in recent years, this has come to be known as "alignment." The diagrams of the curriculum process are meant to indicate the necessity for alignment if objectives are to be achieved. But, the process begins with statements of purpose (mission statements), more specific aims and objectives, and behavioural outcomes (objectives) (see also Chapter 3). The success of the model also depends on an alignment of the student characteristics at entry (e.g., "A" level examinations in the UK, ACT/SAT Scores in the U.S.). Research over the years has sought to determine the power of entry qualifications, which are essentially tests of academic intelligence, to predict the students who are likely to persist or drop out for one reason or another, including failure in examinations. It might be that an alternative curriculum would have to reconsider how students are selected (see Section 4.6).

The model is well illustrated by the problem of curriculum (syllabus) overload which will be considered in the next section.

2.4   THE CURRICULUM PROCESS AND THE OVERLOADED CURRICULUM

These curriculum models are unusual in that they incorporate the syllabus. They are intended to illustrate that the syllabus should be the outcome of a complex activity involving the declaration of objectives and simultaneous design of assessment and instruction procedures that should cause the objectives to be obtained.

The process may be illustrated by consideration of the student complaint that courses are overloaded. By this they mean that the syllabuses are so detailed that they cannot be covered adequately by the teacher or themselves in the time allowed. This does not mean that they would want the course to be lengthened. Indeed, one study showed that the students did not want the course to be lengthened, it was already long enough. "Both engineering and science seniors described how, after trying to keep up with a pace and load to the detriment of a solid understanding of the work, and always feeling too stressed to appreciate the implications of what they were doing, they had shifted into a more measured time frame. As a matter of survival, and in order to get the most out of their education, they advised younger students to give themselves 'permission' to see their major as more than a four year commitment, and to work at less intense pace" [32, p. 221]. This was written in the context of an investigation that sought to understand why undergraduates in science, maths, and engineering courses switched away from these disciplines to non-science subjects, and the fact that many did not complete these science and technology programmes in four years. Seymour and Hewitt who conducted that study hypothesised "that, on every campus, there are substantial numbers of able students who could be retained in SME majors were appropriate changes made in departmental practices." They recognised that change is made difficult by "deeply socialised attitudes." The views expressed in their report have been largely supported by a study of Hamilton College published in 2014 [33] (see Section 3.7).

While it has been understood for years by those who specialise in student health that universities are places that cause stress, it does not seem to have worried many academics, some of whom consider it to be part of the rite of passage. But, if it had been hidden to the public before, Covid has brought the issue fairly and squarely to public attention. Moreover, there are numerous contributing factors, not least of which are systems of examining and assessment [34, 35] and, surprisingly, residence halls that evidently fail to create a supportive environment [36] (see Section 3.7).

Although the term objective is more often than not associated with outcome, it is useful to consider it in its more general sense, that is, something we want to achieve. For example, if the major aim is to develop skill in the transfer of learning, then it is well understood that successful transfer depends on the learner understanding the "key" concepts and principles required for transfer to be successful. It is less well understood that some concepts require greater attention than others in spite of the ease with which students can misperceive concepts [37].

Indirect support for this proposition comes from R. W. Brown who discussed physiological parameters and learning in engineering education [38]. He suggested that one of the reasons why there is so often a mismatch between teacher expectations and student performance is that teachers may be trying to pump too much information down a band-limited channel in too short a period of time. He recorded that his own lecture notes contain 300 Bits per minute, whereas humans could only absorb about 12 Bits per minute over periods of many hours. This led him to advocate the use of pictures because they are "worth a thousand hours." Citing M. D. Lemonick [39], he pointed out that storage and recall mechanisms are very complex. For example, information presented through writing, speaking, and applying is stored in different parts of the brain. He quoted Lemonick, thus "we think of learning and memory as separate functions: in fact they are not. Both are processes by which we acquire and store new data that makes them retrievable later on." The ability to track our approaches to learning in on-line situations means that we will have a much better idea of how to structure and present material.

To determine whether or not a course would be overloaded the estimates of times to be taken for each instructional procedure required for the learning of a key concept or a higher order thinking skill should summed. If the sum of the periods required to complete all these strategies comes to more than the time allowed for the course, then the course is overloaded. This is irrespective of any overloading caused by home study requirements. Therefore, the instructor should be prepared to reduce the number of key concepts and/or higher order skills taught. This will involve him/her in a ranking exercise.

G. Mansfield who echoed this position in respect of the design of mini-courses wrote, "Try to be as realistic as possible, total up the times for all activities on your mini-course outline. Adjust any item to meet the overall goal within the allotted time trading, deleting activities or even reducing the number of realisable objectives." ... "Alternatively, consider providing more total time for the course.....be brutally honest in your time estimates" [40]. This is why the syllabus (content) has been put at the centre of the curriculum models, because it is the outcome of the curriculum design process, and this it seems is difficult to appreciate, especially for those whose beliefs derive from a scholar-academic ideology [41].

Instructors have to cope with the reality of learning, which is that the rate of internalisation varies considerably among any group of students.

Many years ago, when Jim Stice was advocating the use of objectives, he reported that he had found that having to use objectives helped to distinguish between the essential and what was nice-to-know knowledge. This enabled him to cover a course that had never been fully covered before [42].

These are, of course, American studies but they are reminder that we all hold philosophies (beliefs and attitudes), for the most part untested that dictate how we approach student learning. It is these that have to be changed, if change is to be made.

2.5   THE CURRICULUM AS A MIX OF IDEOLOGIES

M. Morant writing of electronics as an academic subject said, "A clear course philosophy is also a good basis for determining strategic priorities. Higher education is responsible to students, industry and society in general to provide the best possible education with limited time and material resources available. A logical basis is required for determining how to use resources for maximum efficiency." The underlying principles on which this philosophy should be based are that the objectives of higher education should be, "To teach students how to think constructively in their subject.... For vocational courses to develop particular and personal skills needed to start a professional career....to develop student's personality and 'world views' in a well-rounded way" [43].

There is nothing special about this statement other than as a statement of the aims of higher education it precedes discussion of a course in electronics. Most academics would subscribe to this or a similar statement, and it is around such statements that the academic identity is constructed even though the actions taken by individuals may be contrary to the achievement of those aims. It is the value system of higher education. When we join a department or agree to teach a course we buy into the system of values associated with it. Over and above this operational (working) philosophy we bring with us belief systems related on the one hand to the nature of knowledge and, on the other hand, to how it is learnt. They impact on how we teach.

The most prevalent system in engineering seems to be what John Eggleston has called a received curriculum, but, as we have seen, there are other belief systems [44]. As is shown in these texts, it is unlikely that any one ideology will support the range of activities required by today's curriculum. While the received curriculum or scholar academic ideology, as it has also been called, supports the lecture-driven approach, the progressive ideologies that follow in the footsteps of pragmatists support inquiry-based learning and project and problem-based work. Thus, the curriculum is likely to be the result of a mix of ideologies.

Ideologies stem from what we want to achieve; thus, curriculum design begins with a statement of aims which will be informed by our beliefs and prejudices. Hence, the need to be able to defend those beliefs: your philosophy.

2.6   THE AIMS OF EDUCATION

Variants of the curriculum process shown in Exhibit 2.1 often replace the title "Knowledge, Learning Skills, and Values" with "Aims and Objectives." The terms "Goal" and "Purpose" are also used as substitutes for "Aims." Colin Wringe notes that philosophers such as John Dewey, Bertrand Russell, and Albert North Whitehead blurred the distinction between aims and ideals [45]. It is clear that both Dewey and Whitehead drew into their thinking the realm of the psychology of learning. While it seems that many people say that the purpose of education is "to educate the whole person" that is based on an implicit idea that may not be the same as the next persons. So it is that this goal has to be described. In Chapter 1 we found a suitable description in the work of John Macmurray. An equally valid view is to be found in Newman's *Idea of a University* [46].

The term "objectives" which has been substituted by "outcomes" is a statement of what a learner will achieve as a result of the curriculum process. It is reductionist and the outcomes of the curriculum tend to be expressed in many statements of what a person "will be able to do." For example, the ability to solve problems is broken down into the sub-abilities that are thought to contribute problem solving. In this language the "aim" of developing skills in the sub-abilities is problem solving. Some writers call the development of problem solving an "objective" as distinct from the sub-abilities which they call "behavioural objectives" [47]. This distinction seems to have disappeared from the current vocabulary but it was popular in lesson planning in high schools.

What is clear is that aims are not objectives. Wringe argues that "the essential logical feature of aims is that by contrast with objectives, they are of an open-ended, on-going kind" (p. 14). In consequence they may never be completely achieved. For example, in engineering education there is continuing discussion about equality and diversity that is accompanied by a continuous flow of developments towards achieving those goals.

Unlike Wringe I am less worried about the mix of ideals in the work of Dewey and White-head because they lead from fundamental epistemologies to a practical curriculum, and in the case of higher education, especially to John Henry Newman's *"Idea of a University."*

Aims in this view are agreed statements of purpose. They function at several levels. There are those that lead to the structure and funding of educational systems. There are those that derive from these first order aims and impact on the curriculum, and at a third order there are those that relate to particular learning activities. At the first order level the prevailing view in some countries in the western world, particularly the UK and the United States is that the function of higher education is to provide graduates that are likely to add value to economic growth. In consequence priority is given to funding that is directed at STEM subjects. In the UK, students are actively discouraged by some politicians from studying subjects that will not bring sufficient earnings to enable them to pay off their student loans. At the second level, that of the curriculum, industrialists complain that graduates are inadequately prepared for immediate work in industry. This relates to a more general debate about the relationship between the academic (theoretical) and vocational (practical). At the third level, different ideologies lead to different approaches to teaching as is illustrated by the debate between the constructivists and realists.

There are many sources of educational aims—political (including legislated aims), eco-nomic (costs; labour force), social (social inequalities, social mobility), ideological (epistemolog-ical, ontological), psychological (development, intelligence) to name but five, each with several categories, that impact on the curriculum.

When he addressed the 1960 summer institute for young engineering educators at Penn State University, Ralph Tyler identified three sources of objectives, the first of which, as we have seen, tends to be forgotten, namely "information about the students to be taught" [48]. In the diagram of the curriculum process this is labelled "entering characteristics." Similarly, less attention is paid to "the conditions and problems of contemporary life that indicate demands that society is making upon young people and adults and opportunities available to them." This remains a pressing problem and relates to all the areas listed above. His third suggestion "comes from the deliberations of specialists in each of the subject matter fields who suggest what contributions they think their subject is able to make to the education of young people."

In contrast to Wringe, Tyler uses the term objectives and embraces aims within it. Some will be more important than others, whereas others might call for contradictory patterns of behaviour on the part of those called on to implement them in lectures, laboratories, and studios. Thus, the outcomes of the process have to be related to its aims, if there is to be both academic and social learning.

Because these debates generally yield more aims and objectives than the system, institution, or college can hope to attain (an axiom that applies as much to the generation of outcomes) it is necessary that at all levels a few significant and consistent aims are selected. The activity that creates such lists is called *screening*. Philosophy provides statements of values against which aims and objectives may be evaluated. Although written about schools, Furst's example is pertinent to engineering education [49]. It relates to Schiro's curriculum ideologies as shown in the brackets. It reads:

"Should the institution prepare students to accept the present social order or should it encourage them to try and improve society? (Social reconstruction ideology). If an institution accepts the first alternative it will emphasise conformity to the present way of living. Such an institution is likely emphasise mastery of fairly stable and well-organised bodies of subject matter on the assumption that there are certain essentials all students should accept." (Scholar-Academic ideology—received curriculum). But it does not end there, because it also supports a rigid system of grading that reinforces the meritocracy. A system that was established to create mobility and abolish paternalism has ended up doing the opposite. While several writers have been good at analysing it, few have any solutions to the problem [50–52]. One thing is clear there can be no change without a fundamental recasting of the aims of education, and that would mean challenging the utilitarian philosophy that currently governs, which so long as the public perceives it to provide social mobility, is likely to be difficult. As Peter Mandler wrote, "No-one has ever lost money on betting on widening participation, and a continued upgrading of educational qualifications, far into the future." But, he adds, "On the other hand I am not advising you to bet on it either. After all, history is full of surprises" [53, p. 215]. Could it be that Daniel Susskind's forecast of the loss of jobs will be the trigger? (see Section 4.4).

To continue, if, as Furst wrote, "An institution accepts the second alternative, it will emphasise sensitivity to social problems and proposing solutions, independence and self-direction, freedom of enquiry, and self-discipline" [54] (social reconstruction ideology).

Screening dictates that the teacher should have a defensible philosophy of education. But it does not end there for it is necessary to screen the proposed list against what is known from psychology about development and learning, moreover there is more than one theory of learning, and they need to be distinguished from theories of instruction. For example, if it is accepted that Perry's view that curriculum and instruction impede student development [55], then steps would have to be taken to ensure that the instructional processes enhance development: assessment should ensure that this takes place [56]. Just as a teacher should have a defensible philosophy of engineering education so too they should have a defensible theory of learning and development.

In this, writers experience the activity of screening is not a linear process. It is complex and results in a detailed curriculum framework. In this text the guiding philosophy is that of the "whole person," as expressed in Chapter 1, and in the philosophies of Macmurray and Newman.

In Chapter 3, consideration of the costs of higher education leads to questions about the purposes of higher education, and the quality of teaching. Since governments fund higher education for its potential contribution to the economy the impact of technology on the present and future structure of workforce is considered in Chapter 4. It is followed by a more specific study in Chapter 5 of education in service of employment and, in particular, the skills that the various partners in higher education believe a new graduate should possess. Drawing from the findings of the study thus far, a model of a curriculum that faces the future is derived in Chapters 6 and 7. In this way the process of constructing a curriculum against present day issues, in particular technological literacy, is described.

2.7   NOTES AND REFERENCES

[1] See Chapter 1 note [10] for a list of the categories in the cognitive domain. The second volume published in 1964 was on the Affective Domain whose categories will be found in note [7, Chapter 5]. 17

[2] Anderson, L. W. and Krathwohl, D. R. et al. (Eds.) (2001). *A Taxonomy for Learning, Teaching, and Assessing. A Revision of Bloom's Taxonomy of Educational Objectives*. New York, Longman. 17

[3] Otter, S. (1991). *What can Graduates do? A Consultative Document*. Sheffield, Employment Department. Unit for Continuing Adult Education. 17

[4] Otter, S. (1992). *Learning Outcomes in Higher Education*. London, HMSO for the Employment Department. 17

[5] Jessup, G. (1991). *Outcomes. NVQ's and the Emerging Model of Education and Training*. London, Falmer. 17

[6] Carter, R. G. (1985). Taxonomy of educational objectives for professional education. *Studies in Higher Education*, 10(2):135–149. 17

[7] These points are discussed in detail in Chapters 3 and 5 of Heywood, J. (2016). *The Assessment of Learning in Engineering Education. Practice and Policy*. Hoboken, NJ, IEEE/Wiley. 17

[8] O'Toole, A. M. (1999). An evaluation of Jerome Bruner's contribution to education at the end of the twentieth century. M. Ed. thesis. Dublin, University of Dublin. 18

[9] Korte, R., Mina, M., and Frezza, S. T. (2019). *Engineering Epistemology. Pragmatism and the Generalized Empirical Method*. San Rafeal, CA, Morgan & Claypool. 18

[10] Dias, P. (2013). The engineer's identity crisis. *Homo Faber* or *Homo Sapiens*? In D. P. Michelfelder, N. McCarthy, and Goldberg, D. E. (Eds.). *Philosophy and Engineering: Reflections on Practice, Principles, and Process*. Dordrecht, Springer. 18

[11] "Working (operational) philosophy" is analogous with the "working capital" required to run a firm. By operational philosophy is meant the value system that drives a particular curriculum, syllabus, course, or teaching session. (i) It is the personal motivation of individuals that sustains them or drives them to change. While its significance does not seem to be widely understood Kazar and Gehrke have demonstrated its value in achieving educational reform. (ii) They evaluated four large communities of transformation concerned with the reform of STEM education. 19, 31, 32

A community of practice is a group of people who share a concern or a passion for something they do and learn how to do it as they interact regularly. Communities of Transformation are variants that are found in the STEM reform area. "The defining feature of these newly identified entities is their focus on exploring philosophically, in deep fundamental ways, how science is taught. This can lead to more substantive changes that have the potential to address the problems described in national reports around underrepresentation of women and undeserved minorities, persistence rates, and success among students. These communities of transformation create innovative spaces that have the potential to shift institutional and disciplinary norms. We identify how these differ from more traditional professional development models, including campus-based professional development and disciplinary meetings." These communities did not exist within organisations from which they could draw resources. This promotes problems of sustainability for such entities. This distinguishes them from communities of practice. They exhibited a compelling philosophy, living integration of the philosophy to create a new world of practice, and a network of peers to break the isolation and brainstorm revising practices.

(i) Heywood, J. (2005). *Engineering Education. Research and Development in Curriculum and Instruction*, pages 55–56, Hoboken, NJ, IEEE Press/Wiley.

(ii) Kezar, A. and Gehrke, S. (2015). *Communities of Transformation and their Work Scaling STEM Reform*. Pullias Center for Higher Education, University of Southern California, CA.

[12] Goldman, S. L. (2004). Why we need a philosophy of engineering: A work in progress. *Interdisciplinary Science Reviews*, 29(2):163–176. 19

[13] *Ibid*. 19

[14] Koen, B. V. (2003). *Discussion of the Method: Conducting the Engineer's Approach to Problem Solving*. New York, Oxford University Press. 19

[15] Koen, B. V. (2010). Quo vadis, humans? Engineering the survival of the species invan de Poel, I. and Goldberg, D. (Eds.). *Philosophy and Engineering: An Emerging Agenda*. Heidelberg, Springer. 20

[16] Youngman, M. B., Oxtoby, R., Monk, J. D., and Heywood, J. (1978). *Analysing Jobs*. Aldershot, Gower Press. 20

[17] (i) Council of Engineering Institutions (CEI) (1968). *Guidelines for Training Professional Engineers*. Statement no. 6, London, CEI. 20

(ii) Council of Engineering Institutions (1968). *Guidelines for Training Management*. Statement no. 9, London, CEI.

(iii) Engineering Industries Training Board (EITB) (1968). *The Training of Engineers*. Booklet no. 5, Watford, EITB.

(iv) Engineering Industries Training Board (1968). *The Training of Managers*. Booklet no. 6, Watford, EITB.

[18] Trevelyan, J. (2014). *The Making of an Expert Engineer*. Leiden, The Netherlands, CRC Press. 20

[19] *Loc. cit*. Ref. [11, (i), p. 3]. 20

[20] *Ibid*. p. 7. 20

[21] Furst, E. J. (1958). *Constructing Evaluation Instruments*. New York, David Mackay. 20, 33, 34

[22] Kerr, J. F. (1963). *Practical Work in School Science*. Leicester. 20

[23] *Loc. cit*. Ref. [11, (i), p. 8]. 21

[24] Pears, A., Daniels, M., Nylén, A., and McDermott, R. (2019). When is quality assurance a constructive force in engineering education? In press. *ASEE/IEEE Proc. Frontiers in Education Conference*. Cincinnati, OH. 21

[25] Borrego, M. (2008). Creating a culture of assessment in engineering education. *ASEE/IEEE Proc. Frontiers in Education Conference*, S2B:1–6. 22

[26] Marton, F. and Säljö, R. (1976). On qualitative 1. Outcomes and process 2. Outcomes as a function of the learner's perception of task. *British Journal of Educational Psychology*, 46:4–11 and 46:115–127. 22

[27] The examination boards that were responsible that setting the examinations—known as "A" levels—that judged whether students were suitable for entry to University. 22

[28] For a description of "A" level engineering Science see Chapter 3 of Heywood, J. (2016). *The Assessment of Learning in Engineering Education. Practice and Policy*. Hoboken, NJ. 22

[29] *Loc. cit.* Ref. [11, p. 284]. 23

[30] Heywood, J., McGuiness, S., and Murphy, D. (1980). The public examinations evaluation project. *The Final Report*. Dublin, University of Dublin, School of Education. 23

[31] Heywood, J. (1992). Student teachers as researchers of instruction in the classroom in J. H. C. Vonk and H. J. van Helden (Eds.). *New Prospects for Teacher Education Europe*. Brussels, Association for Teacher Education in Europe. 23

[32] Seymour, E. and Hewitt, N. M. (1997). *Talking About Leaving. Why Undergraduates Leave the Science*. Boulder, CO, Westview Press. 24

[33] Chambliss, D. F. and Takacs, C. G. (2014). *How College Works*. Cambridge MA, Harvard University Press. 24

[34] Malleson, N. (1966). *A Handbook of British Student Health Services*. London, Pitman. 24

[35] Ryle, A. (1969). *Student Casualties*. London, Allen Lane. 24

[36] Marriott, J. (2021). *The Times*, September. 24

[37] Clement, J. (1981). Solving problems with formulas: Some limitations. *Engineering Education*, 25:150–162. 24

[38] Brown, R. W. (2000). Physiological parameters and learning. *ASEE/IEEE Proc. Frontiers in Education Conference*, 2(S2B):7–11. 25, 34

[39] Lemonick, M. D. (1995). Glimpses of the mind. *Time*, pages 52–60, July 31. 25

[40] Mansfield, G. (1979). Designing your own min course. *Engineering Education*, 70(2):205–207. 25

On the assumption that the syllabus must remain the same Felder, Stice, and Rugarcia (2000) (i) cite Felder and Brent who argued that much of the material that is used in lectures can be assigned to handouts or even coursework pack. The handouts should have spaces for the students to fill in missing steps. I have used this technique, but at the appropriate point in a lecture told the students what to put in the space. The blanks were always for key concepts or important principles (ii). In another course I provided a self-study guide designed to accompany the lectures.

(i) Felder, R. M., Stice, J., and Rugarcia, A. (2000). The future of engineering education VI. Making reform happen. *Chemical Engineering Education*, 34(3):209–215.

(ii) Heywood, J. and Montagu-Pollock, H. (1977). *Science for Arts Students. A Case Study in Curriculum Development.* London, Society for Research into Higher Education.

[41] Heywood, J. (2018). *Empowering Professional Teaching in Engineering. Sustaining the Scholarship of Teaching.* Morgan & Claypool Publishers. 25

[42] Stice, J. E. (1976). A first step toward improved teaching. *Engineering Education*, 70(2):175–180. 25

[43] Morant, M. J. (1993). Electronics as an academic subject. *International Journal of Electrical Engineering Education*, pages 110–123. 26

[44] Eggleston, J. (1977). *The Sociology of the School Curriculum.* London, Routledge and Kegan Paul. 26

[45] Wringe, C. (1988). *Understanding Educational Aims.* London, Unwin Hyman. 26

[46] Newman, J. H. (1947 edition). *The Idea of a University.* London, Longmans, Green with an introduction by C. F. Harrold. 27

[47] Cohen, L. and Mannion, L. (1977). *A Guide to Teaching Practice*, 1st ed., London, Methuen. 27

[48] Tyler, R. W. (1960). Notes for "Conducting Classes to Optimize Learning Summer Institute on Effective Teaching for Young Engineering Teachers," September 1960 at the Pennsylvania State University. 27

[49] *Loc. cit.* Ref. [21]. 28

[50] d'Ancona, M. (2021). *Identity, Ignorance, Innovation.* London, Hodder and Stoughton. 28

[51] Markovits, D. (2019). *The Meritocracy Trap*. London, Allen Lane. 28

[52] Sandel, M. J. (2020). *The Tyranny of Merit. What's Become of the Common Good*. London, Allen Lane. 28

[53] Mandler, P. (2020). *The Crisis of the Meritocracy. Britain's Transition to Mass Education since the Second World War*. Oxford, Oxford University Press. 28

[54] *Loc. cit.* Ref. [21]. 28

[55] Perry, W. G. (1970). *Intellectual and Ethical Development in College Years: A Scheme*. New York, Holt, Rinehart and Winston. 29

Perry proposed that a student entering university progresses from a simple dualistic view of life and knowledge, in which absolute answers exist for everything, to relativism, in which knowledge and value judgements are seen as relativistic (position 5). Following this, an adjustment is made to the relativistic world, and in the last three stages the student begins to experience commitment, and at first make an initial commitment in an area such as career selection, values, and religious belief. In the ninth position, commitment becomes "an on-going unfolding activity through which the student's life style is expressed." It is argued that university teaching tends to reinforce dualism, and seldom helps students develop beyond position 5. (See Ref. [38, pages 109–121], and for an alternative model focused on the development of reflective judgement.)

[56] Pavelich, N. J. and Moore, W. S. (1996). Measuring the effect of experiential education using the Perry model. *Journal of Engineering Education*, 85:287–292. 29

CHAPTER 3

In Search of Aims

3.1   INTRODUCTION

Given that engineering education is a sub-system of higher education it is with the aims of higher education that the screening has to begin. There is no shortage of material that bears on the politics of higher education. For example, the Irish edition of the *London Times*, on May 30, 2019, reported a review (Augar report) of student fees and funding in the UK's higher and further education sectors which led to a bunch of letters in the following editions. Five days later, Melanie Phillips, a right-of-centre columnist writing in the same newspaper, suggested that the famous, for some notorious, British Prime Minister Margaret Thatcher, while achieving much that was good, did Britain harm because she did not believe there was such thing as society. Melanie Phillips wrote, "Her outlook was constructed almost entirely around a utilitarian economic model of human behaviour in which the individual was first and foremost a consumer. Market forces would determine which demands would be the fittest to survive. Choice was elevated to a sacred principle. Value was judged solely by measurable outcomes such as productivity, efficiency or cost effectiveness. Public service was replaced by managerialism. Intangible community bonds that connected us to each other such as trust, duty, or collective solidarity were ignored."

The present day universities are a product of that view. Their purpose is to serve economic growth, serving the community is quite a different matter. However, in this model of higher education, *in extremis*, the student does not have a choice of subject, rather it is argued that they should only study subjects that will bring an economic return sufficient for them to repay the loans they have had to take out to pay for their tuition and living. In this model the graduate is a product of the process of higher education which is utilitarian, and managed like a business. Covid brought home to the public an awareness that universities had not caught up with the idea that their chief customers were their students, and that there was a growing public literature which showed an enormous amount of dissatisfaction among students with the service they received (see Section 3.7). Fees, it was shown, were being used to top up research, and the value of a degree was diminishing because of grade inflation.

Perhaps the most significant indicator that universities have rejected the idea of the university that was developed in the 19th century are the successful attempts by faculty and students to impose restrictions on free speech in the very institutions that claim to be bastions of free speech. Thus, Melanie Phillips points to a culture war between the left running amok with their ideological inquisitions, and the hyper individualism in the social sphere that was reinforced by Mrs. Thatcher in the economic sphere.

We read that "at St. Andrews the induction asks students to agree with statements including 'acknowledging your personal guilt is a useful start point in overcoming unconscious bias.' Those who tick 'disagree' are marked incorrect and too many wrong answers mean they have failed the module and must retake it" [1]. According to this report St. Andrews will not allow students to matriculate unless they agree with certain statements related to compulsory modules in sustainability, diversity, consent, and good academic practice.

Universities should surely be places that help students negotiate differing ideologies as their minds develop. One of the things that we know from neuroscience is that the brain develops well into an individual's twenties. Until then it is plastic, a fact that has enormous implications for the organisation of education, and in particular the curriculum.

There would not seem to be much hope that the universities would meet this obligation if another article, published a couple of days after Melanie Phillips commentary, also in *The Times*, is to be believed. James Marriott's comment was titled "Soulless universities eat away at mental health" with the sub-title "Many students are left isolated and desperate as rapid expansion crushes community of spirit." Not a sign of the acceptance of the "whole person" aim. The activity of screening asks what remedies can be put in place if this view is unacceptable, and what role individual lecturers (professors) have in its remediation.

On June 5th, the BBC Radio 4 "Today" programme was hosted by the University of York. The listener heard it argued that thousands more students should be admitted to university in order to bridge the participation gap between lower earning (working class) families and those of middle and upper earning classes. The universities were being told that they should be agents of social mobility. Yet another function of universities.

Screening would examine the concept of mobility and find it to be a more difficult concept than one might imagine. For example, it would be necessary to distinguish between absolute and relative rates of mobility. Policy makers will certainly need specialist advice, and when they do, they will find that data presented by Erzsébet Bukodi and John Goldthorpe of Nuffield College, Oxford leads to the conclusion that "Educational policy as the primary means of creating more equal mobility chances has, in the light of historical evidence, to be seen as misguided" [2]. But, educational policy makers will also find some of Bukodi and Goldthorpe's recommendations of significance, in particular that the focus "should be on policies for economic and social development of a purposive kind" [3].

Writing about "Education isn't Enough" in the July, 2019 issue of *The Atlantic*, a U.S. millionaire said that he had given millions to education as a philanthropic cause. But he had now come to the conclusion that he had been wrong to believe that education could improve the lot of American children especially those in low-income and working-class communities. At the same time he thought there were serious flaws in the system that needed to be repaired, "We have confused a symptom—educational inequality—with the underlying disease: economic inequality. Schooling may boost the prospects of individual workers, but it doesn't change the core problem, which is that the bottom 90% is divvying up a shrinking share of national wealth. Fixing that problem will require wealthy people not only to give more, but take less."

Participating in that BBC "Today" programme at the University of York was Sir Simon Jenkins a previous editor of *The Times* now a columnist with *The Guardian*. A long-time critic of the English Education system, he told the audience that the universities were a magnificent middle-class confidence trick. He said their practices were archaic: he had yet to find anyone who could explain why degree programs had to be of three years' duration, or why the period of study in each of those years was so short. In England, students on leaving high school for university will be expected to attend full time and complete the bachelor's degree in three years. In Scotland, students take four years. The differences arise from relative differences in school attainment at the point of entry arising from different curriculum structures.

The screening activity cannot avoid seeking answers to his question "what are the purposes of higher education?," or, examining them in the light of the term (semester) structure of higher education.

The programme concluded with a comment by a former Secretary of State for Education Justine Greening who was very critical of the Augur review. She argued that higher education should be free at the point of delivery and paid for by a graduate tax rather than a student loan, thereby making the point that there are alternative systems to the student loan system too which politicians and the British press are wedded.

These illustrations show just how complex the issues are that confront anyone who would try to make radical changes to the higher education system. It has seemed to be relatively immune to disruption. The question is whether or not changes in technology coupled with the increasing cost of higher education will be sufficiently disruptive to bring about the changes that seem to be required. Does one try and analyse the system with a view to affirming its aims or deriving new ones or, does one start with a clear goal to be reached and work out how to get there? It is certain that whatever proposals are made those who make them will have to show that they can be financed.

The examination of the tuition fee debate in the UK and the U.S. that follows does not lead to new aims but it reinforces the need to answer Sir Simon Jenkins question what is higher education, and whom is it for? It does show that some educators are thinking about shortened programmes with the same content.

3.2   TUITION FEES IN THE UNITED STATES

In the U.S. debate about the financing of higher education, and in particular, student loans has led to fundamental questions about "who" and "what" is higher education for?

As a result of the banking crisis that occurred a decade ago, which most of us seem to have forgotten, the term "bubble" has once again became common parlance. Some writers think that higher education is either in a bubble or about to become a bubble. In the U.S., tuition fees had risen by 274% in the years between 1990 and 2009 which was more than the price of any basket of goods or services. State funding at non-profit, public, higher-education institutions has fallen dramatically during the last decade and prices have risen at twice the rate for four-year private institutions [4].

Surprisingly, given that the labour compensation share in net income remained relatively static throughout the whole period, these rises did not cause a major debate about tuition fees, given the importance attached by families to higher education [5]. To an extent this continues to be true, and again surprisingly, in *"Game of Loans,"* Beth Akers and Matthew Chingos argue that typical borrowers of student loans face affordable debt burdens. That is, in the U.S. they are able to repay them irrespective of the level of debt because of what in England is called the graduate premium (see Section 3.4) [6]. But, while repayment may be feasible, they point out that the fees market is dysfunctional since competition among colleges is having the effect of driving tuition fees up rather than down. Yet, if higher education is a market this makes sense. Those colleges that are perceived to have status will be in a position to up their fees, and those that wish to up their status will take fee levels to be some indicator of status. Colleges chase the richer students. This they can do by providing luxury facilities. The search for status functions in parallel with "structural curriculum drift," as for example every college must have a business school (with which to attract overseas students) [7]. It also functions in parallel with an increasing emphasis by universities on research, part of which is paid for by student fees. The implications are profound for secondary (post-elementary) education (see Section 5.5).

Overall, tuition costs increased most for those students who could afford them least [8]. No wonder that in the 2016 U.S. Presidential election young people supported Bernie Sanders and caused Hilary Clinton to state her intention of making in-state tuition free at public colleges and universities for all Americans whose families earn less than $125,000 a year. This was similar to what occurred in the UK in 2017 when student loans became a major issue. Students came out in force to vote for left-wing Labour Party leader Jeremy Corbin to the total surprise of the establishment as he proposed the abolition of student fees! The debate continued into the 2020 U.S. Presidential election.

It is also argued that the fees dilemma arises from a particular philosophy, namely, that the purpose of college education is to serve the economic good. Since "free market" thinking dominates economic and political thinking, it is held that all educational institutions should be run like businesses and compete with each other. In theory, that would be all well and good but for the financialization that led to the collapse of the world economy in 2008.

Foroohar [9] whose book *Makers and Takers* gives an explanation of the origins of that crisis paraphrased the views of Adair Turner [9, (a)] to explain financialization. She writes, "Turner is saying that rather than funding the new ideas, and projects to create jobs and raise wages, finance has shifted its attention to securitizing existing assets (like homes, stocks, bonds, and such), turning them into tradeable products that can be spliced, and diced and sold as many times as possible—that is until things blow up, as they did in 2008. Turner estimates that a mere

15% of all financial flows now go into projects in the real economy." Banks changed their role and created economies based on debt. From the perspective of this study it is not only the lack of investment but of the parallel impact of developments in technology on employment that are important.

Immediately, however, it is with the possibility that the financialization of education may have the same effect. Foroohar also takes this position. She writes, "Cryn Johannsen lays out in *Solving the Student Loan Crisis* that 'students…are defined as consumers seeking out person-alised education and training that will make them marketable.' It is a philosophy that disconnects higher education from its value as a public good, and the development of the individual. It marks a major change of purpose. American higher education was never completely devoid of mercan-tilism (for-profit business and trade schools have been around since the 19th century), and its virtually never been free; but payment for it was in the past split more evenly between families, the government, and philanthropy; and, civic benefits were as highly valued as the economic ones (which, crucially, were seen as accruing to the nation, rather than just the individual)" [10]. All of this raises fundamental questions for the screening exercise, forcing its first activity to be a consideration of the purpose of higher education.

3.3 WHAT IS HIGHER EDUCATION FOR?

Foroohar points out that in the U.S. the possession of a graduate qualification is no longer a ticket to social mobility, indeed, it "can result in downward mobility" [11]. Robert Reich, public servant and economist, takes a similar view. "The demand for well-educated workers in the United States seems to have peaked around 2000 and then fallen as the supply of well-educated workers has continued to grow [...] since 2000 the vast majority of college graduates have experienced little or no income gains at all [...] To state it in another way, while college education has become a pre-requisite for joining the middle class, it is no longer a sure means of gaining ground once admitted to it" [12, p. 209], [13].

Foroohar points out that between now and 2024 America will create 14,000,000 jobs. They will nearly all require at least a two-year associate's degree, a consequence of which is that community colleges should become "the new high schools"—"a basic necessity for every American."

Support for this argument comes from data about the U.S. workforce [14]. About a third of new jobs up to 2020 will require no more than an associate degree, but they will require something more than a high school diploma. These jobs have come to be called "New Middle Skilled" and contrary to Reich, it may be argued from an earnings perspective that many of these jobs will bring workers into the middle class [15, 16]. (See also Section 2.5.)

Foroohar also argues that too many degrees from for-profit colleges in the U.S. are mean-ingless: her two examples are sports marketing and business administration! There are also those who argue that the quality of the for-profit college degree programmes are poor [17].

She argues that the higher-education system should be reformed, and that it should be recognised that the arts are as important as STEM subjects, hence the acronym STEAM [18]. This point was argued by this writer at the American Society for Engineering Education in response to a report to the National Governors Association recommending a refocussing away from the liberal arts towards vocational studies in four-year public colleges [19, 20].

At the far end of the spectrum are futurists like Israeli philosopher Yuval Noah Harari [21] or Max Tegmark [22] who contemplate a world in which AI and robotics have put us all out of work. One consequence of the mass unemployment envisaged is that we would all have to receive a basic income from the State. In these circumstances would higher education be redundant, or would it be a preparation for leisure, or the place where one finds one's personal identity? [23] If the latter, how would it be financed? That raises an additional question to Foroohar's "for whom is higher education for?" And, that is "What is higher education for?"

3.4   TUITION FEES IN THE UK: THE GRADUATE PREMIUM

Charging fees for University Education is a recent phenomenon, in England, Northern Ireland, and Wales where most universities receive public funding. The Scottish Government declined to follow suit and does not charge fees. Fees in the other countries of the UK were charged from 2004 when a cap was set at £3,000. The cap is now over £9,000, soon to be increased. It has become universally charged throughout the university system. The legislation received support from both sides of parliament, but not the Liberal Party. Since 2010 there has been declining support for tuition fees and according to Andrew Rawnsley, Andrew Adonis (Labor) the self-described moving force behind the 2004 legislation says that "fees have become so politically diseased they should be abolished entirely" [24].

In response, it is argued that the size of the tuition fee has not put off the supply of applicants from socially disadvantaged groups. Indeed, their numbers have increased. However, a report from the Social Market Foundation questioned data from the Higher Education Statistics Agency [25]. The data presented by the Foundation shows that many of the disadvantaged groups targeted through widening access are also the groups who are most likely to drop out. The report does not consider student loans but it does say that financial constraints contribute to drop out, and Andrew Rawnsley makes the link with loans. It should also be noted that another study reports an earnings gap of 10% (at the median) between graduates from higher- and lower-income households [26]. But the same study offers statistics to show "that a degree offers a pathway to relatively high earnings for a large subset of graduates from across institutions."

The principle behind the student loan is the graduate premium, that is, the difference between a graduate's earnings over a lifetime compared with a non-graduates. The Minister (Margaret Hodge) who made the proposal argued that it was approximately £400,000. But one research has put it at probably less than £100,000. This gives an annual premium of just £2,222 per year before tax. Its author, Kemp-King, wrote that that is not enough to cover the interest on the loan [27]. It is important to note that the premium not only varies with institution attended but with subject as well. Medicine and dentistry are at the top of the scale, and subjects like sports science and the creative arts, at the bottom. It has also been suggested that the impact of technology on graduate jobs is likely to be profound which would affect the premium (see Chapter 4).

When the system was legislated it was agreed that students would not start repaying the loan until they were earning £21,000. It was believed that most graduates would earn above that sum, but that has not turned out to be the case. It is expected that many more graduates than was anticipated will not complete the repayments on their loans. The figure for those graduating in 2015 was put at 70%.

An important finding of the Akers and Chingos investigation in the U.S. was that only a quarter of first-year students can predict their debt load within 10% of the correct amount. This was put down in part to the complex deals they are asked to negotiate which change from year to year. In the UK, Ian King wrote "most graduates and their families still confuse the size of their debt with the far lower payments on it they will in reality face. Hence, all the discontent as graduates fret about their ability to obtain a mortgage" [28]. Worrying about financing a mortgage is probably less important than worrying about financing a pension. Put the two together and the debt load begins to grow. Be that as it may discontent is not likely to go away as students find that those who own the loan book can alter the rates of interest they have to pay on loans.

3.5   THE IMPACT OF UNIVERSITIES AND THE FINANCIAL SYSTEM ON FAMILIES

Attendance at university in both the UK and the U.S. is now seen by middle class parents as a right-of-passage for their children. There is, therefore, a pressure on families to push their children through universities, and to find such money as they can to support them while at university. Nevertheless, very many students find themselves having to take out substantial loans which can as we have just seen impede their life's chances. A recent research by Caitlin Zaloom in the U.S. brings a quite different focus to the loans debate because it looks at the impact that borrowing large sums of money has on middle class families, and the moral conflicts that it creates [29]. The task she set herself was immensely difficult because of the "silence" of middle class families who deal with their difficulties in the privacy of their own homes and prevent neighbour intrusion. "The collective refusal to speak openly about money has also impoverished our vocabulary for describing the trade-offs and dilemmas that the financial system imposes on families, and undermined our capacity to understand how being indebted shapes families lives" (p. 29). Viewed in this way universities are part of the financial system.

Zaloom points out in contrast to Akers and Chingos who regard debt as a transaction between present and future, that debt requires that money is redirected towards financial institutions for many years and away from the pursuit of life's aspirations, as for example, those students in the UK who are unable to take on a mortgage. For many students debt is an inhibitor to achieving that highly prized goal of higher education of placing students on a route that enables them to continue to explore their potential. Universities and the financial system conspire to shut many students down.

It follows that comparisons of different methods of funding higher education can only be evaluated when the moral compass in which they function is taken into account. Curriculum design cannot ignore this dimension. For this reason alone curriculum designers should consider the merits of less costly alternative structures of higher education.

3.6   AN ALTERNATIVE STRUCTURE FOR UK DEGREE PROGRAMMES

Surprisingly, Sir Simon Jenkins did not talk about two-year programmes when he questioned the credibility of the UK's three-year bachelor's degree programmes, yet he had mentioned them in his column a couple of years previously. One private university (Buckingham) had from its founding run a two-year programme for the bachelor's qualification. It compressed the three-year programme into two.

In 2014, a former Minister for the Universities, John Denham saw such compressed programmes (39 weeks per year) as a means of reducing tuition and residence costs by 20%. Subsequently the Government decided to allow such accelerated courses. However, universities that do so will be allowed to charge a higher annual fee which might defeat the objective of reducing the costs of courses to students [30]. A small number of universities that were created in 1992 are offering such programmes.

In an address to the Royal Society of Arts and Manufactures, John Denham also considered the fact that employers say graduates lack employability. He argued employees should be subsidised by their employers through university on two-year programmes. Employers and universities would work together to design the right course.

The idea of two-year degrees (not compressed) is not new academically or politically. In 1968, Sir Brian Pippard, the Cavendish Professor of Experimental Physics at Cambridge, proposed a reorganisation of the structure of university degrees and teaching. In respect of physics he suggested a two year plus two year system. The first two years would be designed for those who would require only a general understanding of physics, whereas the full four year course would be for those seeking a career in physics. The same might be argued about all the sciences and engineering. Can the humanities escape the same argument?

Ten years later in a report on higher education known as the Brown Paper, its author Gordon Oakes, the Minister for Higher Education, proposed a two-year course for the abler student [31]. Although nothing came of the proposal it should be noted that the Treasury expressed a preference for less able students "because they could absorb less" [32].

The issue of two-year programmes has been given vent again in recent weeks. Sir Simon Jenkins' wrote in *the Guardian* lamenting the wiping out of the Polytechnics in the 1990s while at the same time accusing universities of being "bastions of privilege." This brought a response, also in the *Guardian*, from the Master of Churchill College Cambridge that could have come straight out of the Percy Committee report on Higher Technological Education of 1945 [33]. It had expressed satisfaction with universities as vehicles for preparing scientists and technologists for research, but dissatisfaction with the technical college course intended to produce scientists and technologists for industry. It caused the establishment of Colleges of Advanced Technology for the purpose of educating graduates for industry by means of four-year sandwich courses.

In essence, The Master of Churchill beginning with the premise that universities provided an academic education, for four years in the sciences and engineering whereas the polytechnics of the past provided part time training for a different kind of qualification, urged, "that in parallel with any introduction of such two-year courses, institutions continuing to offer three- or four-year degree should be recognised as providing education directed towards a distinctly different end-point. Perhaps this would require a differentiation in institution type to re-invent polytechnics, by another name if necessary," which is exactly what the Percy committee debated. This was of course the idea behind the Diploma in Technology initiative of the 1950s. It did not work (see Sections 3.1 and 5.1). It also assumes that employers can predict their needs in the long run. There is no evidence that they can. Subsequent sections of this study wrestle with this problem.

3.7 THE QUALITY OF TEACHING AND THE HIDDEN CURRICULUM

As we have seen, the high costs of fees have raised questions about the quality of teaching in both the UK and the U.S. which some claim is very poor. The response of the UK Government has been to introduce a starred system for rating the quality of teaching that will impact on the fees a university is allowed to charge.

There have been a series of articles in the quality newspapers about poor teaching which would not inspire applicants or their parents. For example:

"Students are getting a third class education" [34].

"Students taught one to one for only 26 hours in entire degree" [35].

The first article was a collection of comments from students in different universities who complained about the tuition they received. The second, is more important: it reports on a survey undertaken by two economists who investigated the contact hours between academic staff and students. Their purpose was to develop a quantifiable measure of the teaching resource provided by a university that could also provide a rough cost of tuition. It confirms the conclusion of Jenni Russell, the author of the first article, that students were being short-changed. They found that universities that charged more did not offer any more teaching. She wrote "unfortunately the ideology of independent learning the Russell group has developed results in students being taught in large classes with minimal feedback. This inevitably has consequences for the amount of work they undertake." Johnes and Johnes [36] (cited by Holmes and Mayhew [37]) "reported that the cost of providing undergraduate courses in the UK is significantly less than the fees being, and that undergraduate fees are being used to subsidize research."

That ideology is in stark contrast to the student-oriented view of the university propounded and followed by Newman in the 19th century. In this respect the findings of a ten year study of Hamilton College are of interest [38]. This small private liberal arts college was the subject of a ten year study intended to find out "How College Works." The authors Daniel Chambliss and Christopher Takacs asked the question, "Can students get more out of college without spending more money?" This meant that they had to examine how the quality of college education could be improved without additional cost. The answer is surprising. They believed it could.

They found that the single most important thing in the quality of a student's education was to do with the way a college is organised to help the students with their relationships, and that went for the classroom experience as well. Relationships, "are the necessary precondition, the daily motivator, and the most valuable outcome. A student must have friends, needs good teachers, and benefits from mentors. A student must have friends, or she will drop out physically or withdraw mentally. When good teachers are encountered early, they legitimize academic involvement, while poor teachers destroy the reputation of departments and even entire institutions. Mentors, we found, can be valuable and even life changing……relationships are important because they raise or suppress the motivation to learn, a good college fosters the relationships that lead to motivation."

Newman also considered that a successful education depended on relationships. "When," he wrote, "a multitude of young men, keen, open-hearted, sympathetic and observant, as young men are, come together and freely mix with each other, they are sure to learn one from another, even if there be no one to teach them; the conversation of all is a series of lectures to each, and they gain for themselves new ideas, and views, fresh matters of thought, and distinct principles for judging and acting, day by day" [39].

It was this view of university education that evidently influenced head teachers, when in the 1960s they compared universities with the Colleges of Advanced Technology, to the disadvantage of the latter. If it was true then, it is patently not now, and the consequences can be profound. Alice Thomson wrote in 2018, "A friends son left a London university this year and transferred to York after the student in the next room committed suicide. 'None of us even knew his name.' London was overwhelming. I was commuting in from my halls for 45 minutes every day and sitting on my bed alone in the evening. I worried that it could be me next" [40]. As we have seen, three years later, Marriott was to make a similar point about residence (see Section 3.1).

Marriott raised the question why is it necessary for people to move away from their local university. If universities don't perform this function then a student's costs can be reduced. But Newman offers a rebuttal thus: "A parallel teaching is necessary for our social being, and it is secured by a large school or a college; and this effect may be fairly called in its own department an enlargement of mind. It is seeing the world on a small field with little trouble; for the pupils or students come from very different places, and with widely different notions, and there is much to generalise, much to adjust, much to eliminate, there are inter-relations to be defined, and conventional rules to be established, in the process, by which the whole assemblage is moulded together, and gains one tone and one character" [41]. It follows that teaching and residence are interdependent since together they achieve the goal of enlarging the mind, and where students are day students some mechanism has to be found to achieve this effect. Further to this argument is the view that a large department in a large university is no different to a small college, and given this view of a university has the same obligations.

Both the Hamilton College Study and Newman's idea demonstrate that the curriculum as conceived here, extends far beyond the classroom; and that this "hidden" curriculum is as important for learning as the formal curriculum, as is the affective to the cognitive [42].

3.8   GRADE INFLATION AND THE USEFULNESS OF SOME DEGREE PROGRAMMES

Criticisms of education are made at all levels. University education in the UK is criticised because in the last 20 years there has been strong grade inflation [43]. In an earlier period when the problem of grade deflation was severe in the U.S. it was explained that it was a function of the view that society took of grades in that particular period [44]. Milton and his colleagues also argued that teachers necessarily adapt to changing circumstances.

In the UK, grade inflation has had the predictable effect of lowering the ceiling of jobs for which a degree is necessary. To put the picture in another way, many graduates are employed in jobs for which they are overqualified [45]. But, their pay is more likely to be related to the job than to the degree. Therefore, there is an argument that students should be steered away from taking degrees that will result in lower earnings. "Graduates who study the creative arts, for example, tend to earn less and so over time we might be concerned that these shifts will bring down the graduate earnings premium. What is not clear is the reason for these changes in subject mix" [46]. Apart from the fact that students want to study these subjects Holmes and Mayhew suggested that some universities offer more lower-cost courses since fees do not vary by subject. This, they argue, should be taken into account when reckoning the public subsidy for education.

Clearly, they have been somewhat defeated because students do not follow the rational expectations model that economists have of the person. Student interests do not necessarily correlate with their academic ability. Moreover, the information available to them is asymmetric and governed by all sorts of factors outside of their academic profile. Envisaging a future that is clouded with uncertainty is very difficult, and student is just as likely to use rules of thumb that have been successful in the past, as not.

The model supports those who argue that, in any event, many undergraduates are not suited to higher education, but it does not support the view that they would necessarily be better off in vocational courses as some commentators, political advisers, and distinguished academics recommend they do. In any event, the middle classes in the UK and U.S. are unlikely to encourage their children to take vocational courses.

3.9   THE MIDDLE CLASSES AND VOCATIONAL COURSES

A major problem, in the UK at least, is one that will not go away. It is that of status. Vocational courses and the jobs associated with them are perceived by the middle classes to have lower status. A university degree is perceived as necessary to maintain status. Moreover, the middle classes do not necessarily perceive that a vocational qualification will fulfil the promise that those who advocate them suggest they will. They might never have heard of T. H. Marshall, but they certainly understand the point he was making, that once a person has embarked on a vocational route, especially for a semi-profession, there may be no ladder out (see Chapter 7 [34]).

According to Zaloom the same is true of the U.S.: "In recent years, a chorus of politicians, policy experts, and economically minded columnists have located the value of college in preparing young people for jobs"… in particular they are recommended to STEM fields "rather than allowing them, let alone encouraging them, to devote themselves to pursuits seen as less pragmatic and the development of skills portrayed as in less demand" (p. 164). Which is the mirror of the British picture.

Zaloom makes the important point that when the pressures are on to do degree programs that will have high earning power by doing the jobs that corporations wish students to do, these advocates are recommending what "amounts to little more than higher level vocational education for the middle class" (p. 165). She notes that there is evidence that the idea that a liberal arts education will not get them good jobs is spurious [47].

Related to those who advocate "Career and Technical Education" (CTE) she notes, as did Marshall, that CTE may train persons for jobs that may be eliminated in the future which is the thrust of Susskind's argument (see Section 4.4). Zaloom adds that "Both the advocacy for CTE and the argument for college as preparation for middle class yeomanship conflict with the traditional American understanding of the value of higher education, and citizens' rights to it."

3.10   QUESTIONS ABOUT THE PURPOSES OF HIGHER EDUCATION: TECHNOLOGY AND LIBERAL EDUCATION

So, taken together with complaints about teaching, we come back to the same question "what is university education for?" Students and their parents and probably most taxpayers are unlikely to be satisfied with the response given by the head of a humanities faculty in a Russell group university to questions about student complaints about teaching. She replied to Andrew Rawnsley in

The Observer with a sigh, "It's our fault. We've not being good at making the students understand that teaching is not what we are here for." So what are universities for?

In the absence of an answer to this question the system is unlikely to change, as it has done in the past, that is, in small steps that maintain some continuity with the past while at the same time allowing business methods to dominate policy making. Given all the factors, but particularly that of student loans at a time when parental incomes are roughly static, the likelihood of the time span of programmes being generally reduced (e.g., two years in the UK, and three years in the U.S.) is great, although there will be much resistance to it. Furthermore, perceptions about what is required for the world of work will focus on trying to make individuals take vocational courses that provide a higher remuneration than they would otherwise get from pursuing degree programmes with less rewarding outcomes (see below). This assumes that things will remain pretty much as they are: technologies will put people out of work but new technologies will provide an equivalent number of jobs to those that have been lost. But, this is an assumption that is being challenged by several authorities including leaders in technology (see Chapter 4). If they are correct it will require a fundamental re-appraisal of the purposes of higher education.

In the context of the U.S., Zaloom returns to the traditional understanding that Americans have of higher education which derives from the political philosophy of John Dewey. He argued, she writes "that education should teach students how to fashion novel habits, dispositions, and institutions, serving to advance democracy as circumstances evolve." Education could not be reduced to simple preparation for jobs, he argued without damaging both students and their country and the world. He acknowledged that vocational training could be part of the educational framework, as long as it included social, political, and moral dimensions. But he cautioned "The kind of vocational education in which I am interested is not one which will 'adapt' workers to existing industrial regime; I am not sufficiently in love with the regime for that" [48]. Instead, he advocated "for a vocational education that would enable workers to transform the industrial system, a goal that embraced the rapidly evolving circumstances of the early 20th century and saw the possibility of justice for them" (p. 169). It would seem that specific training belongs to the work place; at least, that is the position taken here.

From the perspective of curriculum design in the 21st century, Zaloom's account is at the level of generality that seems to reinforce the division between the academic and the practical, yet it is impossible to argue that a person has been liberally educated who has no acquaintance with technology both the experiential and principle. Furthermore, if we follow Newman's logic the mere acquaintance with a range of subjects is not a liberal education. They have to help a person view a problem from all angles.

Given that education is a preparation for life, and that life is embedded in technology it should respond to those generic skills that belong to technology as well as those that belong more generally to living and learning. The technological dimension of liberal education might be described as technological literacy which necessarily engages the liberal and the vocational. The curriculum designer's task is two-fold: first, to define technological literacy, and second to integrate the skill and knowledge obtained into an integrated system of learning that will be the end product for some students, and the beginning for others. Irrespective of capability all students require a minimum competency in technological literacy, a model curriculum for which is discussed in Chapter 7.

3.11   CONCLUSION

Apart from issues of inequality, arguments about student loans in both the UK and the U.S. have led to questions about the structure of higher education, quality of teaching, content of the curriculum, and irrelevance of certain degrees to the economy which students persist in taking. They lead to questions about the purposes of higher education—"For whom?" and "for what?" The current answer is the utilitarian business model-based system of higher education we have now. Some authorities, however, question this model and in the U.S. suggest that there should be a return to the educational and political philosophy of John Dewey which has many similarities with the concept of a university promoted by John Henry Newman. It is argued here that for a liberal education to be liberal it has to embrace technology and a particular model of how this might be achieved is given in Chapter 7.

In the next chapter these issues are considered from the perspective of the impact of technology on the structure of the workforce.

3.12   NOTES AND REFERENCES

[1] Woolcock, N. (2021). Pass bias test to enter St. Andrews. *The Times*, October 1. 36, 49

[2] Bukodi, E. and Goldthorpe, J. H. (2019). *Social Mobility and Education in Britain: Research, Politics, and Policy*. Cambridge, Cambridge University Press. 36

[3] *Ibid*, p. 223. "Educational expansion and reform over the last century or more *have* widened opportunity in the sense that more individuals of all social origins alike have been able more fully to realise their academic potential. And it is on making further progress in this regard-on reducing the significant wastage of talent that still occurs-that those who work within the educational system should be required, and allowed to concentrate, rather than having imposed on them, under an unduly instrumental view of education, the leading role in overcoming of inequality in the wider sense, the main sources of which lie outside of educational institutions. Efforts can still be made to deal with these problems, as far as possible, through other forms of policy-ones that need to be aimed in one way or another at reducing the effects of inequalities of condition. But insofar as social mobility *per se* is to remain a concern—rather than just a convenient topic for political rhetoric—the main focus should be on policies for economic and social development of a purposive kind. That is, for development directed towards the creation of a technologically and economically more efficient and also more humane society that would lead through

changing demand conditions, to a steadily increasing number of men and women, of all social origins, being able to move into class positions in which they could enjoy economic well-being, security and stability and the prospect of advancement over the course of their working lives." 36

[4] Akers, B. and Chingos, M. M. (2016). *Game of Loans. The Rhetoric and Reality of Student Debt*. Princeton, NJ, Princeton University Press. 38

[5] Erixon, F. and Weigel, B. (2016). *The Innovation Illusion. How so Little is Created by so Many Working so Hard*. New Haven, CT, Yale University Press, p. 205 cites the U.S. Bureau of Labor Statistics and U.S. Bureau of Economic Analysis and includes graph. 38

[6] *Loc. cit.* Ref. [1]. 38, 50

[7] Curriculum drift is a term developed in the UK to describe how the curriculum focus moved away from a focus on the production of engineers for industry to mirror that offered in university engineering and technological studies by Colleges of Advanced Technology in the late 1950s and early 1960s. In general, educational institutions seek higher status by mirroring the institution above them in the status tree. 38, 50

[8] Goldrick-Rab, S. (2016). *Paying the Price: College Costs, Financial Aid, and the Betrayal of the American Dream*. Chicago, University of Chicago Press cited by Foroohar, R. (2016) in "How the financing of colleges may lead to disaster." *New York Review of Books*, October 13. See also review by Laurie Taylor from the perspective of a UK academic https://www.timeshighereducation.com/books/review-paying-the-price. 38

[9] Foroohar, R. (2016). *Makers and Takers. The Rise of Finance and the Fall of American Business*. New York, Crown Business/Penguin Random House. 38

Describes a change in the role of banks from funding new investment to lending against existing assets thereby creating economies based on debt. For a UK view (a). For another U.S. view (b).

(a) Turner, A. (2016). *Between Debt and the Devil. Money, Credit, and Fixing Global Finance*. Princeton, NJ, Princeton University Press.

(b) Reich, R. B. (2016). *Saving Capitalism for the Many, Not the Few*. New York, Alfred A. Knopf.

(c) For a mapping of the system with an illustration of the Bear Stearns collapse see Bookstaber, R. and Kenett, D. Y. (2016). *Looking Deeper, Seeing More: A multilayer Map of the Financial System. OFR Brief Series*. June 16–July 14, 2016. Washington DC, Office of Financial Research.

[10] Foroohar also takes this position in a book review. How the financing of colleges may lead to disaster. New York Review of Books, October 13, 2016. See Chapter 3 of O'Neil, C. (2016), *Weapons of Math Destruction. How Big Data Increases Inequality and Threatens Democracy*. London, Penguin Books. Describes the impact of rating scales on the development of Universities in the U.S. For a description of similar developments in the UK see Hale, T. and Vina, G. (2016). University challenge. The race for money students and status. FT magazine June 27, 2017. https://next.ft.com/content/c662168a-38c-11e6-a780-b48ed7b6126f. 39

[11] Foroohar, R. (2016). How the financing of colleges may lead to disaster. *New York Review of Books*, October 13, 2016. 39

[12] *Loc. cit.* Ref. [6, (b)]. 39

[13] Autor, D. H. (2015). Polyani's Paradox and the shape of employment growth. *Re-Evaluating Labor Market Dynamics*, pages 129–79, Kansas City, MO, Federal Reserve Bank of Kansas City. 39

[14] Career Vision. Opportunities abound in middle skill jobs. https://careervision.org/opportinities-abound-middle-skill-jobs, August 14, 2017. Provides links to U.S. Department of Labor and Bureau of Labor Statistics. 39

[15] Carnevale, A. P., Rose, S. J., and Cheah, B. (2011). *The College Payoff, Occupations, Lifetime Earnings*. Washington, DC, Georgetown University Center on Education and the Workforce. 39

[16] Goodwin, B. (2012). Research says don't overlook middle skill jobs. *Educational Leadership*, 89(7):86–87. 39

[17] See Chapter 4 of O'Neil, C. *Loc. cit.* note [7]. 39

[18] *Loc. cit.* Ref. [7]. 40

[19] Sparks, E. and Waits, M. J. (2011). *Degrees for What Jobs? Raising Expectations for Universities and Colleges in a Global Economy*. Washington, DC, National Governors Association. 40

[20] Heywood, J. (2012). Education at the crossroads: Implications for educational policy makers. *Distinguished Lecture at Annual Conference of the American Society for Engineering Education*. 40

[21] Harari, Y. N. (2017). *Homo Deus. A Brief History of Tomorrow*. Dvir Publishing. 40

[22] Tegmark, M. (2017). *Life 3.0. Being Human in an Age of Artificial Intelligence*. 40

[23] Heywood, J. (2014). Engineering education in search of divergent vision. Who am I? Who are you? Where are we going? *Keynote Address 44th Annual ASEE/IEEE Frontiers in Education Conference*, Madrid. 40

"There is one other dimension of change I wish draw to your attention, and that is 'identity.' It is a term that is very familiar which was brought to public attention by the German born American psychoanalyst Erik Eriksen (a). He related it to the search for identity that goes on between the ages of 12 and 18 during which time the adolescent tries to find his or her identity. We might put it in terms of a search to answer the question 'Who am I?' Eriksen took the view that if a person does not solve that problem that person is likely to experience role confusion. Given the many pressures on youngsters in this age group a few of them are likely to become very confused. Very few will avoid some confusion, and they will all change to some extent or another: and that will be the sign of an individual's development. This is what I meant by internal change at the beginning. It is brought about by the interactions we have with other people. Today we are concerned with the relationships that a person has with the education community on the one hand and on the other hand the professional community and the confusions that exist between the two."

"The difficulty with Eriksen's stage of identity is that it can so easily give the impression, unintended I am sure, that a person reaches maturity at the age of eighteen (b). My own view is somewhat different. It is my submission that we continually search for an identity throughout life and that we experience many confusions between work and life as well as within work and life (c). Consequently, all change involves a capacity to deal with ourselves as we construct, maintain and develop our identity. We go along with that which leaves our identity unaffected. We drive for change if we believe we will find our identity. All change involves a capacity to deal with ourselves as we construct, maintain and develop our identity. We resist that which we think will shatter our identity. All change involves changes in attitudes, beliefs, and values. Our conceptual understanding continually develops. We are continually seeking the answer to the question 'Who am I?' In its restricted form, for example in relation to the group we are in, and in its general form in relation to the 'life,' we live. It is the confusions that are caused by the profession of engineering education that are of considerable importance to us."

"All this may be said of groups (d) [...]"

(a) Eriksen, E. (1950). *Childhood and Society*. New York, Norton. Parents should allow children to explore and not try to get them to conform to their views. This exploration seeks answers to the questions who am I? And, how do I fit in? See also, The problem of ego identity (1956). *Journal of the American Psychoanalytic Association*, 4:56–121. "Identity means the partly conscious, largely unconscious sense of who one is, both as a person and a contributor to society," cited by Hoare, C. (2006). Work as the catalyst of adult development and learning in Hoare, C. (Ed.). *Handbook of Adult Development and Learning*. Oxford, Oxford University Press, p. 348. On the same page Hoare writes [...] "the original identity construct, as it was defined and described by Erik Erikson from within his and U.S. societal lens, incorporates a decided vocational commitment."

(b) I have heard distinguished university educators argue that because children and adolescents have studied a wide range of subjects in school that they have received a liberal education. This justifies the four or three years study in one or two subjects that follows school in Irish universities.

(c) Support for this view will be found in Hoare *ibid*. She writes "both personal dimensions-identity and adult personality-evolve more or less so, as one learns and grows. Each dimension is partly conscious and partly unconscious. When adults are deeply work engaged, they function in a time-out-of-mind zone, rarely surfacing to ponder (if in fact they consciously can) their sense of self and its vital constituents. Clearly other personal attributes are also important to occupational conditions" (p. 347). Some workers either cannot or will not develop and continue to learn.

(d) Korte, R. (2007). A review of social identity theory with implications for training and development. *Journal of European Industrial Training*, 31(3):166–180. "Social identity is one lens through which individuals view their jobs, responsibilities, organizations and even the dynamics of work (e.g., causal attributions). Therefore, social identity becomes an important lens through which people perceive new information, attribute causes, make meaning, and choose to undertake new learning. Without addressing the identity factors stemming from group membership, the success of typical training efforts may fail to realize their promise of improving individual and organizational performance."

(e) Avent, R. in a special report in *The Observer* 9/10/2016. Welcome to a world without work writes. "It (work) is also a source of personal identity. It helps give structure to our days and our lives. It offers the possibility of personal fulfilment that comes from being of use to others and it is a critical part of the glue that holds society together and smooths its operation. Over the last generation work has become ever less effective at performing these roles. That in turn, has placed pressure on government services and budgets, contributing to a more poisonous and less generous politics."

[24] Rawnsley, A. (2017). You don't need a double first to see university funding is in chaos. *The Observer*, p. 35, 09:07:2017. 40

[25] Social Market Foundation (2017). *On Course for Success? Student Retention at University*. London, Social Market Foundation. 40

[26] Britton, J., Dearden, L., Shephard, N., and Vignoles, A. (2016). How English domiciled graduate earnings vary with gender, institution attended, subject, and socio-economic background. *IFS Working Paper W16/06*. London, Institute of Fiscal Studies. 40

[27] Kemp-King, S. (2016). *The Graduate Premium: Mann, Myth, or Plain Mis-selling?* London, Intergenerational Foundation. 41

[28] King, I. (2017). The lessons from rising student loan debt have not been learnt. *The Times*, p. 36, 25:07:2017. 41

[29] Zaloom, C. (2019). *Indebted. How Families make College Work at Any Cost*. Princeton, NJ, Princeton University Press. 41

Moral mandates embedded in the financing of students through college (Zaloom pages 16–20). (1) That children obtain a degree in the first place. (2) The family should be nuclear. Assessment is based on this idealized model in the U.S. (3) That parents take charge of college finances. Parents hold a moral obligation to pay. "If they can't come up with the funds for college by being fiscally prudent and investing, then they must be willing to take on costly debt, and they must therefore also accept whatever future constraints on their spending and life choices cost of repayment imposes." (4) It is up to the student to get those jobs for which they are qualified that will enable them to pay back their loans. It follows that "that the value of higher education is primarily financial rather than about open futures." […] "As for parents, in addition to getting and staying married, the morally tinged assumption of the student finance complex is that they will have no problem paying back their loans if they manage their careers well, no matter how the conditions of their fields might evolve."

[30] Two-year university courses come a step closer. 24:02:2017. https://www.gov.uk/government/publications/amendments-tabled-ahead-of-lords-report-stage 42, 54

[31] Department of Education and Science (1978). *Higher Education into the 1990s*. London, HMSO. 42

[32] Stevens, R. (2004). *University to Uni. The Politics of Higher Education Since 1944*, pages 42–43, London, Politico (Methuen). 42, 63

[33] *Higher Technological Education*. London, HMSO. Report of a Committee chaired by Lord Eustace Percy. 43, 63

[34] Russell, J. (2017). Students are getting a third class education. Many universities don't care about teaching quality and undergraduates don't realise it until it's too late. *The Times*, p. 18, 23:03:2017. 43, 46, 63

[35] Bennett, R. (education editor). Students taught one to one for only 26 hours in entire degree. *The Times*, p. 18, 31:07:2017. This article is about a study of contact hours in UK universities. The authors of that report have a separate article in the same issue. Huxley, G. and Peacey, M. Tuition fees must be linked to quality. 43, 54

Russell group. A grouping of UK universities not including Oxford and Cambridge intended to form an élite.

[36] Johnes, G. and Johnes, J. (2016). Costs, efficiency, and economies of scale and scope in English higher education. *Oxford Review of Economic Policy*, 32(4):596–614. 43

[37] Holmes, C. and Mayhew, K. (2016). The economics of higher education. *Oxford Review of Economic Policy*, 32(4):475–496. 43

[38] Chambliss, D. and Takacs, C. (2014). *How College Works*. Cambridge. MA, Harvard University Press. 44

[39] Newman, J. H. (1842–1946 (edited) Harrold, C. F.). The idea of a university. *Defined and Illustrated*, pages 129–130, London, Longmans Green. 44, 54

[40] Thomsom, A. (2018). Universities treat students like cash machines. *The Times*, June 6. 44

[41] *Loc. cit.* Ref. [39]. 45

[42] (i) Lin, H. (1979). The hidden curriculum of introductory physics curriculum. *Engineering Education*, 1:36–40. 45

(ii) Lynch, K. (1989). *The Hidden Curriculum*. London, Falmer Press.

[43] *Loc. cit.* Ref. [30]. The debate about the number of firsts awarded was renewed in July 2019. See *The Times*, July 12, 2019. 45

[44] Milton, O., Pollio, H. R., and Eison, J. A. (1986). *Making Sense of College Grades*. San Francisco, CA, Josei-Bass. The period investigated was 1950–1965. 45

[45] CIPD (2015). Over-qualification and skills mismatch in the graduate labour market. *Policy Report*. London, Chartered Institute of Personnel Development. 45

An EU analysis that shows that apart from five countries the UK has a higher underutilisation rate than the other members of the EU 27. The authors note that much research in this area confuses because it conflates two issues—whether a degree is necessary to get a job, and whether it is needed to do the job.

[46] *Loc. cit.* Ref. [35]. 45

[47] Deming, D. J. (2017). The value of soft skills in the labor market. National Bureau of Economic Research. https://www.nber.org/reporter/2017number4/deming.html 46

[48] Roth, M. (2012). John Dewey's vision of learning as freedom. *New York Times*, September 5. https://www.nytimes.com/2012/09/06/opinion/john-deweys-vision-of-learning-as-freedom.html 47

CHAPTER 4

Technology and the Changing Structure of the Workforce

4.1 IMPACT OF TECHNOLOGICAL WORKFORCE FORECASTING ON EDUCATIONAL POLICY

Workforce forecasting is notoriously difficult, and distinguishing myth from fact is sometimes problematic as Michael Teitelbaum has shown [1]. For example, the apparently quite simple task of categorising a particular "job" can be very difficult. Because the assumptions made by educators, industrialists and researchers may differ the task of interpretation may be as difficult as biblical exegesis!

In the UK there has been a fairly consistent belief during the last 70 years that there is a shortage of qualified scientists and technologists [2]. Estimates made of the numbers of qualified scientists and technologists since 1945 have helped determine future policy in higher technological education. Except for one, these reports held that there was a shortage of such personnel. The exception in 1961 caused much controversy. It said that supply and demand would be in balance by 1965 [3]. The next report in 1963 focussed on shortages, particularly of technologists [4].

The Committee on Scientific Manpower doubted whether employers' statements about their future requirements could be regarded as fully valid. They thought that employers of mechanical engineers should be employing more qualified staff if they were to expand in the next decade. Similar debates have taken place in the U.S., often accompanied by employer lobbying for funding or visas to offset shortages. They often win [5].

In the U.S., Charette used a report from the Georgetown University Centre for Education and the Workforce to illustrate one of the difficulties of workforce forecasting, in this case unforeseen events [6]. The report predicted 2.4 million STEM job openings in the U.S. between 2008 and 2018 with 1.1 million newly created jobs, and the rest to replace workers who retire or move to non-STEM fields; they concluded that there would be roughly 277,000 STEM vacancies per year. Charette pointed out that the study did not fully take into account the effects of the downturn that occurred with the Great Recession. The jobs increase forecast for 2010 did not happen, instead there was a loss of 370,000 science and engineering jobs.

A heading for an article on the front page of the *Wall Street Journal* of October 13, 2016 read, "Tech Boom Creates Too Few Jobs" with the sub-heading, "Dashed Employment Promises of the 1990s Fuel Donald Trump's Political Rise." Among several items in a long article we read that the, "photo-sharing service Instagram had 13 employees when it was acquired for $1 billion by Facebook in 2012."

According to Charette, about 15 million U.S. residents hold at least a bachelor's degree in a STEM discipline, but three-fourths of them—11.4 million—work outside of STEM.

Charette found that there was "an extraordinary amount of inconsistency" among the hundreds of reports, articles, and papers published during the last 60 years. This is consistent with this writer's finding. Yet, as Teitelbaum showed, politicians have consistently responded positively to views that the U.S. is short of scientists and engineers [7]. But Charette argued that this response caused money to be diverted into solving this problem rather than solving the problem of shortage of STEM knowledge that exists in the population as a whole. "We should be figuring out how to make children literate in the sciences, technology and the arts to give them the best foundation to pursue a career and transition to new ones" [8] which has profound implications for the design of the curriculum, and more especially the continuity between school and higher education. Charette did not consider the problem of specific shortages.

4.2   SPECIFIC SHORTAGES

The phrase "specific shortages" has at least two meanings. The first applies specifically in the U.S. and relates specifically to "speciality" or "guest" workers. These are workers who have been brought into the U.S. on H-1B visas. These visas were instigated in response to successful lobbying by employer led groups. They were "justified on grounds that employers requiring scientists, engineers, computer and IT workers were facing debilitating 'shortages' of qualified hires and therefore impeding their ability to compete internationally" [9]. A person seeking such a visa had to possess a bachelor's degree. The legislation came in at a time when the booms that were used to justify these visas came to an end!

In relation to the workforce Teitelbaum identified five rounds of alarm/boom/bust since World War II. Most lasted for a period of between 10 and 20 years. It is difficult to infer from Teitelbaum's data that there was a general shortage of qualified STEM workers [10].

The other meaning of "specific shortage" is not very different. It also relates to speciality shortages; for example, in 2011 the Confederation of British Industry (CBI) reported that 40% of companies had difficulties recruiting people with science, technology, engineering, and maths skills. But it did not say at what level these skills were wanted [11]. Other data reveals specific shortages [12, 13].

A striking example of a specific shortage was that of the resurging nuclear power industry in the U.S. "The persistent demand for nuclear power, coupled with mounting concern about safety, has exposed a dearth of advanced training programs in the increasingly complex skills required. During the three-decade hiatus in nuclear plant construction following the Three Mile Island accident in 1979, many universities phased out their nuclear engineering schools or merged them into other programs. Now, demand for trained personnel is expected to rise. According to the Nuclear Energy Institute's 2010 Work Force Report, nearly 38% of workers in the U.S. nuclear industry will be eligible to retire in the next five years. To maintain the current workforce the industry will need to hire 25,000 more workers by 2015. The U.S. Bureau of Labour Statistics projects an 11% growth in the need for nuclear engineers in the period up to 2018" [14].

A parallel example comes from the UK where in respect of a large underground rail development in London (Crossrail) the contractor reported that while it would require 1000 tunnellers, there were only 500 tunnellers in the whole of the UK, and their average age was 55 years. Unfortunately it did not say what qualifications were required [15].

More generally, in the U.S. some evidence on the demand for personnel seems to suggest that the pattern of demand is changing at all levels of skill including professional work, as the next section on engineers suggests.

4.3   CHANGING PATTERNS IN EMPLOYMENT PROSPECTS IN ENGINEERING: TOWARDS AN ALTERNATIVE MODEL OF HIGHER EDUCATION

Much of the data used in current commentaries was derived several years ago because that is the way data is collected (e.g., census data is usually collected at the end or beginning of a decade).

If it is possible to extrapolate from the experience of Silicon Valley then the demand for technological manpower is declining irrespective of specific shortages. The U.S. Bureau of Labour Statistics recorded for the decade ending 2010 that techno-scientific employment fell by 19%, and that average wages in Silicon Valley fell by 14% [16].

G. Paschal Zachary writing in *IEEE Spectrum* said that often emerging technologies require far fewer workers [17]. The new titans of Silicon Valley employ far fewer workers than the older titans and this is likely to apply equally to their offshore establishments. At the same time some emerging technologies destroy jobs. He also drew attention to the phenomenon of "jobless" innovation. This occurs when an innovation is off-shored to countries where qualified workers are much cheaper to employ.

Related to employment in the software industry was a comment in the "E Mail" column of the November 2011 issue of *ASEE Prism*. It contained an exchange of letters between Professor Allen Plotkin and columnist Vivak Wadwha about an article that Wadwha had written in the September issue of the magazine [18]. He had asked, why should a company pay a 40-year-old engineer a considerable salary if it can get the same job done much more cheaply by an entry-level employee? He said that it was happening in the software industry. "After all the graduate is likely to have more up-to-date skills and work harder."

Wadwha continued "if you listen to the heart-wrenching stories of older engineers" (who have become unemployed) "you learn they have a great many skills, but no one wants to hire them." It seems there may be a serious unemployment problem among middle aged and older engineers in some sectors of the U.S.

While Professor Plotkin questioned whether or not anyone would want to work in an industry that treats its workers in the way described by Wadwha, the way industry employs people is changing dramatically, and some may argue with the Professor for the worse.

For example, an Irish Academic said to this writer that firms had told him that they wanted young graduates who could do the job immediately, but they would only keep them from seven to nine years! A couple of years later Charette went further [19]. He wrote, "The nature of STEM work has also changed dramatically in the past several decades. In engineering, for instance, your job is often no longer linked to a company but to a funded project. Long term employment with a single company has been replaced by a series of *de facto* positions that can quickly end when a project ends, or the market shifts [20]. To be sure, engineers in the 1950s were sometimes laid off during recessions, but they expected to be hired back when the economy picked up. That rarely happens today. And unlike in decades past, employers seldom offer generous education and training benefits to engineers to keep them current, so out-of-work engineers find they quickly become technologically obsolete."

Several fundamental questions arise. It seems clear that industry believes that the costs of education and training should be borne by society in the form of educational institutions (public or private), and the state acting on behalf of society believes that the student should pay the costs of higher education. Two questions arise. First, "What responsibility, if any, should industry have for the education and generic training of the people it employs?" Second, "What responsibility does society acting through the State have for the education and development of the individual beyond their secondary (post-primary) education?"

In relation to Wadwha's remarks and the purposes of higher education, Charette's comment raises the question as to, "how long engineers can make such changes in function before they become unemployable?" Wadwha's response to Professor Plotkin is an indirect answer with a constructive outcome. He cites the metaphor of a roller coaster and argues that universities need to prepare students for that ride so that when the need arises they are able and interested to change jobs.

Irrespective of the extent of middle-aged engineer unemployment or the employment of young engineers for a short period of time it is clear that in terms of knowledge redundancy engineers require spells of "retraining." This term is used in preference to continuous professional development (CPD) since CPD often implies knowledge acquisition in relation to current employment. In some cases engineers may well have to seek employment in activities (jobs) that are at a substantial cognate distance from the jobs they currently have. This implies a need for continuing professional and personal development (CPPD), a term that is not currently in use. It implies that all higher education is in part about providing a base for "permanent" education (a term favoured in the 1960s), and raises questions about the nature (content) of the base education to be offered, as well as the subsequent continuing education to which it is linked.

Heywood suggested that the basic higher education should be of two years duration. He did not envisage three years being compressed into two as proposed by John Denham and operated by institutions such as the University of Buckingham because he thought it would be necessary for individuals to top up their learning at regular intervals throughout their careers (see Chapter 3).

Cheville proposed that this should be funded through insurance. He thought that an insurance policy might be linked to the higher education institution that the person first attended for purpose of credentialing. It was unrealistic, he argued, to place the responsibility for CPD on the individual particularly those individuals working in the GIG economy. He noted that beyond work, as individual's age they do things that society needs such as creating families and seeing they are educated: participating in the work of charities, and contributing to the work of local and central democratic government. They contribute to social as well as economic capital.

He argued that nations have created education systems that are based on hierarchical structures of knowledge that may or may not be "true" or "relevant" in the context of economic and technological change. It forces many young persons into thinking about careers before they need. Add internet learning into the mix and the need for fixed discrete periods of learning disappears.

Cheville contrasted the current mortgaged-based method of financing higher education via student loans with the insurance-based model that is suggested by changes in the workplace and advances in technology, particularly robotics and AI [21]. They create the perception of an investment rather than the yoke of a loan. In the latter, small payments rather than large payments are made. Economically the alternative model has the possibility of bringing in a new pool of talent.

He also believed that the system should provide multiple pathways to success. He pointed out that alternative structures of higher education would inevitably involve changes in attitudes to credentialing [22].

Beyond engineering the impact of technology on the professions has been described by Susskind and Susskind [23].

4.4   THE IMPACT OF TECHNOLOGY ON PROFESSIONAL WORK

The Susskinds point out that, thus far, while the major impact of the digitisation and automation of tasks has been on unskilled, low-skilled workers and low-wage occupations [24], it is now beginning to impact on the jobs done by the lower middle and professional classes. The Susskinds conclude that "increasingly capable non-thinking machines will displace much of the work of human professionals." Young graduates know that this is the case, and some are worried. A survey of 8,000 college educated young professionals from 30 countries revealed that 51% thought they would need to be retrained to stay relevant in their positions, and 40% thought that automation would threaten their current job because there would be less demand for their skills, while 53% said automation would make the workplace more impersonal. But those who had technological skills thought that automation would increase the number of jobs open to them [25].

In contrast, the pessimistic forecast suggests that, "an increasing number of professionals must be absorbed in a decreasing range of types of task (namely, those in which professionals still have the advantage). In short, it will become ever more difficult, as time passes and machines become increasingly capable to ensure that there is enough reasonably-paid employment for professionals" [26, p. 290]. The Susskinds who made this prediction were very clear that they were, "not predicting that the professions will disappear over the next few years. We are looking decades' ahead […] and anticipating incremental transformation and not an overnight revolution" [27, p. 219]. One step in this direction is the development of co-robotics in which individuals and robots work together in situations where the individual holds the comparative advantage. Beyond that are substantial improvements in machine reasoning.

Coincidentally, during the 2020 Pandemic *Covid 19* Daniel Susskind published a follow up to "*The Future of the Professions*" in which he considered more generally technological unemployment and the future [28]. His aim was to ask questions about what happens to us if we have to do less or no work. He maintained the position taken by his father and himself in the earlier publication. "Machines will not do everything in the future, but they will do more. And as they slowly, but relentlessly, take on more and more tasks, human beings will be forced to retreat to an ever-shrinking set of activities. It is unlikely that every person will be able to do what remains to be done; and there is no reason to imagine there will be enough demand for it to employ all those who are indeed able to do it […] And, as we move through the 21st century, the demand for the work of human beings is likely to wither away, gradually. Eventually, what is left will not be enough to provide every-one who wants it with traditional well-paid employment" (p. 5).

Covid 19 has probably made the relatively well-off middle classes aware of the huge inequalities that exist in the labour market, in particular the poor wages paid to health workers on whom we rely to keep hospitals going, and provide homecare to older and infirm people as well as people with a whole range of disabilities. In the UK they get clapped on Thursday nights but will this lead to an increase in their income and a reduction in inequality? Probably not, if the past is anything to go by. The Susskinds' point is that the labour market which is the cause of these inequalities "is already creaking [and that] technological unemployment is simply a more extreme version of that story, but one that ends with some workers receiving nothing at all" (p. 6). The question is, will Covid 19 accelerate these changes?

Susskind and Susskind is about how we should go about resolving the issues that will be promoted by increasing technological unemployment [29]. One of them "the meaning of life" is very much the province of education.

It is concluded, first, that this is not just a problem for the education of professionals but for the lifelong education of everyone. Second, and immediately, the direction of change in professional work and the professions will necessitate changes in the structure of education that professionals receive since, like engineers, they too will need life-long education, and some may need to take non-cognate jobs, as will many of those in the "new middle-skill jobs."

4.5   THE NEW "MIDDLE-SKILL" JOBS

Autor would appear to be much more sanguine because it will take a very long time to overcome the limitations of current technology to accomplish non-routine tasks [30]. He believes that "new middle-skill jobs" will arise that will replace traditional middle-class jobs, and that just as automation in the 1950s did not create unemployment, the same will happen now. But this is to rely on what has happened in the past. The benign view takes the position that although changes in technology are much more rapid than they have been in the past, and although difficulties with machine learning will be overcome more quickly than is currently judged, jobs will continue to be created. Therefore, fears of unemployment are exaggerated.

Autor's [31] benign view of the impact of technology on jobs continues to prevail. Brynjolfsson and McAfee had argued earlier that while jobs are being destroyed new jobs are being created for which new skills are required [32–34]. But, it depends on having the right skill set. Otherwise they caution, "There's never been worse time to be a worker with only 'ordinary' skills and abilities to offer because computers, robots, and other digital technologies are acquiring these skills and abilities at an extraordinary rate" [35, p. 11]. These middle-skill tasks will, to cite Autor, "combine routine technical tasks with non-routine tasks have the comparative advantage: interpersonal interaction, flexibility, adaptability, and problem solving" [36]. Autor cites Holzer to the effect that "new middle-skill jobs" are growing rapidly even in technical production and clerical occupations [37]. The Brookings Institute suggests that augmented reality (AR) can bridge the manufacturing skills gap. They also suggest that it can be used to train workers in new skills which may help reduce the skills gap [38].

Such views are strongly supported by the Organisation for Economic and Cultural development (OECD). They write , "To seize the benefits of technological change, economies need ICT specialists, workers who can code, develop applications, manage networks and manage and analyse Big Data, among other skills. These skills enable innovation in a digital economy to flourish, but also support the infrastructure that firms, governments, commerce, and users rely on [39]. However, besides these experts, digitalisation also calls for all workers to have a relatively high minimum level of ICT skills, even those on low-skilled jobs. For instance, this is the case for blue collar workers in factories that are entirely automated or waiters having to take orders on iPads" [40].

Matthews the editor of *ASEE Prism* wrote "more than jobs is at risk if the United States continues to bleed manufacturing operations [...] loss of manufacturing could also diminish the American capacity for innovation. However, from the pessimism comes hope, even if there is a sting in the tail" [41]. He continued, "Advanced manufacturing, if it succeeds, offers a bright future for engineers [...] Laid-off industrial workers will not fare so well, since part of what makes the new techniques attractive is greater productivity. What will be needed are skilled technicians

Taking care of people
Making computers work
Taking care of business
Building and maintaining our infrastructure
Teaching children
Designing things: solving problems
Keeping businesses running
Selling goods and providing basic services

Exhibit 4.1: Adapted from—Top areas of job growth in the next six years simplified but ranked in order of demand from Waldock, C. Closing America's job gap. U.S. Bureau of Labour Statistics. Cited in Sparks, E. and Waits, M. J. (2011). *Degrees for What Jobs? Raising Expectations for Universities and Colleges in a Global Economy.* Washington, DC. National Governors Association.

with a grounding in math and science" which seems to be somewhat contradictory since that does not presage a need for engineers. Support for this view will be found in Washington State's Assessment of Education Credentials and Employer Needs programme [42]. Eleven Centers of Excellence have been established by that State in two year colleges. The occupations for which skills standards have been developed are all for varying grades of technician and craftsman. In respect of manufacturing, the State of Minnesota has established career and education pathways for a manufacturing and applied manufacturing workers that can bring them as far as middle management on the one hand, and on the other hand, an M.S. degree [43].

According to the President of the Illinois Community College Trustees Association Barbara Oilschlager, 41% of jobs will be at the middle level requiring more education than high school but less than a bachelor's degree [44]. In the UK this would be called technician level education. But, distinctions are made between two levels of technician those requiring one or two years beyond high school and those requiring a basic degree, e.g., engineering technology.

In the U.S. "New Middle-Skill Jobs" that provide the new skill sets have been estimated to provide one third of new job openings between 2010 and 2020 [45]. Many of them are not in technical occupations as Exhibit 4.1 shows. Moreover, some of these jobs produce greater life time earnings than for graduates [46, 47]. While more than a high school diploma is required for many of these jobs, a degree is not, and in consequence some persons can find that the route to these jobs is at the same time a route to social mobility. The most common educational routes to success are via associate degrees and post-secondary certificates in Community Colleges. Thus, Foroohah argues that community colleges need to be revitalized [48]. Similar arguments are made about the technical colleges in England [49]. Industry is also becoming increasingly interested in offering its own courses and certificates, as for example Boeing and

Microsoft. Computer software engineers, aircraft mechanics, electricians are among the high earners [50]. *Career Vision* concludes its pamphlet on these jobs with the statement that, "Since lifelong learning is a characteristic of work today, a middle skill job can also set the stage for further education and even greater career progress" [51, 52].

A finding that distinguishes Ireland from other nations is the youthfulness of its population. Acemoglu and Restrepo show that countries that are ageing rapidly are those that have been at the forefront of the development and utilisation of industrial robots (fully autonomous machines that do not require a human operator), and have also been responsible for the growth these countries have experienced [53]. At the same time in another paper Acemoglu and Restrepo find that thus far the effects of robots in the U.S. "are most pronounced in manufacturing, and in particular in industries most exposed to robots; in routine manual, blue collar, assembly and related occupations, and for workers with less educational qualifications" [54]. They did not find among any of the occupation or education groups positive and offsetting gains in employment. This raises the general question, "If the level of education among these groups is raised will there be sufficient jobs to offset the unemployment created?" A major question for legislators is how to prioritise funding among the different sectors of higher education.

4.6   HYBRID JOBS

The July 2019 issue of *The Atlantic* contains an article by writer Jerry Useem titled "At Work, Expertise is Falling out of Favour." In it he talks about what has been called in the past 'multitasking' where by workers take on more than one task. An early example of this was the replacement of bus conductors by the driver who took the fares as people entered the bus. Useem cites the case of the desk agents at airports who are now being asked to marshal and service aircraft, the persons in these jobs being called "cross utilised agents." He cites a Deloitte consultant who considered that "within the next decade 70–90% of jobs in the future will be hybrid or super." He also cites the 2016 World Economic Forum Report which said that "more than one-third of the desired core skill sets for most occupations" will not have been seen as crucial to the job when the report was published. He reported an IBM person responsible for training, Joanna Daly, who said that the half-life of skills is getting shorter. Apprenticeship programmes for new jobs were now as short as six months.

The thrust of his article is derived from a trip that he took on a littoral U.S. Navy combat ship which had been designed for "minimal manning" with a crew of fifty, each of whom had to multi-task. They had, in his words, to become generalists and undertake a range of skills.

However, the tasks that Useem cited were not in the professional realm, or that of upper technicians. His argument supports the findings of European studies that show that a general education is superior to a vocational education in terms of long term career survival. The findings have considerable implications for the way in which education is structured. They point to the need for lifelong learning, and therefore, the thesis that is emerging from this study so far. In so far as professionals are concerned, it is clear, that many like politicians and civil servants, will need a broad education that takes into account the different mental skills associated with such work [55].

Useem drew attention to the fact that the U.S. Navy had thought that only certain persons would be suitable for these kinds of job. They consulted Professor Zachary Hambrick, a psychologist who undertook some tests on sailors to establish who they might be. He found that those who could multi-task were high on fluid intelligence [56]. This intelligence is different to the crystalized intelligence that is associated to expert performance. Specialization has been to quote Hambrick, "the coin of the realm" but in uncertain environments this is no longer the case. This has considerable implications for education and training, and the development of skill in the transfer of learning which is required in non-cognate situations (see Chapter 6). It also has implications for the selection of students.

The European findings contradict the position taken by policy makers in the UK which may be simplified to say that those who are not academically bright will do better in vocational courses. Since it supports a subject-based model of general education that can be traced back to the beginning of the 20th century education it is reasonable to ask two questions. First, is it or why does it continue to be relevant today? Second, if students in some countries cope with a general education, why is that so many can't in the UK? Could it be that many students require a period of induction which enables them to explore a subject in their own time, and at their own level? Does Whitehead's rhythmic theory of learning suggest a way of structuring the curriculum and learning? (see Chapter 7). Perhaps the way forward is by the inclusion of studies in engineering and technological literacy which might enable more students to utilise their knowledge, which Whitehead believed to be the aim of education. Such utilisation necessarily enlarges the mind, which Newman thought was the goal of education which arises when a person is allowed to explore something from every angle, a skill that is necessary for development.

4.7   TECHNOLOGY PORTENDS A NEW KIND OF DEGREE

As indicated workforce forecasting is notoriously difficult. Nevertheless governments of differing political hues are persuaded by such forecasts as well as the pressure from lobby groups to base their strategies for higher education on such information. However, politicians have not proved themselves to be able to objectively evaluate the data and in consequence produced policy decisions that are muddled.

For example when the conservative government was preparing for the nationalisation of the school curriculum in 1987 the Minister Kenneth Baker produced a curriculum that made no "gesture towards such subjects as economics and business studies, accounting, design and graphics, catering, health and hygiene. Nowhere did Baker relate his preferred subjects to the competitive skills they need for this new world. His was simply a grammar school curriculum enforced on secondary schools by law. There was no choice for parents, teachers, governors-or children" [57]. The Technical Vocational Education Initiative (TVEI) that had shown some promise in this direction was dropped. While technology was included in the curriculum the high status that was promised for this subject by his predecessor Sir Keith Joseph was dropped. This continues to be the case in 2019.

Jenkins describes cabinet meetings where the minutia of the syllabuses in English were discussed such as whether or not art and music should be compulsory. "While nuclear defence might go through the cabinet on the nod, teaching content was a topic on which every minister had a view" [58]. There is no mention of any discussion about technology or its role in education. The ministers were simply replicating the curriculum from which they had benefited. This is not to say that there is a correlation between subjects studied and the decisions made, it is to argue that a cabinet bereft of ministers with science and engineering qualifications is unbalanced and limits the collective wisdom. The law makers in the U.S. seem to be similarly limited. They like the public need to be technologically proficient which suggests, that over and above previous arguments in this chapter, there is a case for a new kind of degree. The problem is that neither the academic or political establishments show any willingness to think outside the box. Fortunately or unfortunately, depending on your perspective, technology is likely to force change through the changes that it forces on the structure of employment.

4.8   CONCLUSIONS

Workforce forecasting is notoriously difficult. Nevertheless governments of different political hues in both the UK and the U.S. are persuaded by such forecasts, and pressure from lobby groups to based their strategies for higher education. Because they are ill equipped academically their decisions are often muddled. It is argued here that beliefs about the shortage of STEM graduates have diverted resources away from the more pressing problem of engaging the school population in general and technological literacy. At the present time, they also have to make decisions based on the optimistic and pessimistic views of the future structure of the workforce that are currently circulating [59]

At the same time, shortages of specific types of technologist will occur which are difficult to forecast. Few occupations will be free from the impact of technology. Many traditional middle-class jobs will be replaced by what have been called in the U.S. "new middle-skill jobs" that will require different skill sets and qualities which do not require a graduate-level programme [60].

Overall the pattern of demand continues to change. During the last decade firms, even firms in silicon-valley have come to employ fewer workers. For reasons of remuneration and rapid changes in knowledge technological firms may prefer to employ younger people and may leave a cadre of unemployed middle-aged engineers. At the same time, many individuals will have to change career paths on more than one occasion during their working lives.

It is clear that some jobs will be lost and not replaced in the geographical areas where they were lost. Displacement will force some workers to seek employment in fields that are non-cognate with the fields they are presently in. Current jargon says that they should be "adaptable." More specifically, they should have skill in non-cognate or cross-domain transfer.

The less certain and the more complex a society becomes the more workers will require personal resilience and the ability to plan their own learning in order to cope with change.

It is the function of the education system as a whole and the sub-system of higher education to provide as much for personal development as it is for the learning that will enable them to adapt. As the patterns of work change so the need for continuing professional development becomes an imperative. A permanent system of education will require changes in the structure in all the sub-systems of education.

4.9   NOTES AND REFERENCES

[1] To a large extent policy has been governed by the regularly reported predictions that there is and will be a shortage of engineers and scientists, and that the pool of students available to pursue these occupations is too small and declining in quality. In both the UK and the U.S. this perception is taken to be correct. It is held that failure to redress this situation will be detrimental to future economic prospects. Much attention has been paid to remedying this shortage particularly by focusing on the supply side of the equation. Michael S. Teitelbaum a Program Director at the Alfred P. Sloan Foundation said at a conference on the U.S. Scientific and Technical Work force "the supposed causes are weaknesses in elementary, secondary, or higher education, inadequate financing of the fields, declining interests in science and engineering among American students, or some combination of these. Thus, it is said that the U.S. must import students, scientists, and engineers from abroad to fill universities and work in the private sector—though even this talent pool may dry up eventually as more foreign nationals find attractive opportunities elsewhere" (a)*. But Teitelbaum went on to argue that such data that was available was weak and often misinterpreted (b)*. There was no evidence for a shortage of qualified personnel and in a submission to a sub-committee of the House of Representatives he said that, "despite lawmakers being told by corporate lobbyists that R&D is being globalized in part due to shortages of scientists in the U.S. no one who has studied the matter with an open mind has been able to find any objective data of such general shortages." He concluded with the controversial view that, "Federal policy encourages an over production of science professionals" (c)*. It has created its own system of vested interests. If the continuing attention to the shortage of students for STEM education is anything to go by this system is alive and well (d)*. Of course it may not be true of other countries (e)*. 57, 71

Much the same applies to the UK In 1961 the *Advisory Council on Scientific Policy* published a controversial report which said that by 1965 the supply and demand for scientific manpower should be in balance which caused a huge row (f)*. In 1963, the Council changed its emphasis and argued that employers of mechanical engineers should be employing more qualified personnel (g)*. Heywood caused a stir at the *1963 Annual Conference of the British Association for Industrial and Commercial Education (BACIE)*, when he argued that on data obtained from employers he found no evidence of a shortage of qualified personnel (h)*.

(a)* Teitelbaum, M. S. (2003). Do we need more scientists? *The Public Interest*, no. 153, National Affairs Inc., Washington DC. He presented this paper at *2007 Conference on the U.S. Scientific and Technical Workforce. Improving Data for Decision Making*. Organized by Rand Science and Technology. The Proceedings were edited by Kelly, T. K., Butz, W. P., Carroll, S., Adamson, D. M., and Bloom, G., pages 11–31. It is interesting to note that forty three years ago John Jewkes in the UK asked a similar question, "how much science?" In his Presidential address to the economics section of the British Association. (*Advancement of Science*, 67, 1960.)

(b)* Lowell, B. Lindsay and Salzman, H. (2007). *Into the Eye of the Storm: Assessing the Evidence on Science and Engineering Education, Quality, and Workforce Demand*. Urban Institute, 48 pages. Also considers that there is no shortage of scientists and engineers and examines in detail the perceptions that have led to the opposite view.

(c)* Cited in *First Bell. Today's Engineering and Technology News* under the heading, Labor researchers tell Congress U.S. not lacking in scientists, engineers. *ASEE*, Washington, DC. See also (a)* *First Bell* 07:06:2011. Some experts say STEM crisis is overblown and contrast with 21:10:2011, demand for STEM skills increasing, study finds. Patel, P. (up dated 2010). "Where the engineering jobs are. The news is good but not great for engineers looking for work in 2010." *IEEE Spectrum*, 03:01:2012.

(d)* For example, (a)* *First Bell* reports on 28:10:2009, High-achievers defect from STEM fields, study finds: 23:05:2011, experts voice concern over high STEM dropout rate: 16:06:2011, training programs offer pointers on incorporating STEM into lessons: 03:02: Technology, engineering overlooked when STEM education discussed, teacher writes (in London *The Times* 01:03:2012 in an article on the importance of science to Britain's recovery no mention is made of engineering): 08:02:2012, Obama to request $80 Million for education funding for training math, science teachers: 13:02:2012, Labor Department official discusses importance of STEM at the University of Dayton.

Ellis, R. A. (2007). *Effects of Recent Revisions in Federal Standard Occupational Classification (SOC Categories of the Employment of STEM Professionals*. New York, Commission on Professionals in Science and Technology. Future of STEM Curricula and Instructional Design. A Blue Sky Workshop. December 1–3, 2009. Center for the Study of the Mathematics Curriculum.

(e)* Blau, J. (updated 19:08:2011). Germany faces shortage of engineers. *IEEE Prism* Downloaded 03:01:2012. Also Schneiderman (2010). Economy and shortages affect European job outlook. The bigger high-tech companies in Europe are recruiting EE's. Talent is in short supply, especially to smaller firms looking for very specific skills. *IEEE Spectrum*, March.

(f)* Advisory Council on Scientific policy. Committee on Scientific Manpower (1961). *The Long Term Demand for Scientific Manpower.* London, HMSO, Cmd 1490.

(g)* Advisory Council on Scientific policy. Committee on Scientific Manpower (1963). *Scientific and Technological Manpower in Great Britain 1962.* London, HMSO, Cmd 2146.

(h)* Heywood, J. (1974). Trends in the supply and demand for qualified manpower in the sixties and seventies. *The Vocational Aspect of Education*, 26(64):65–72.

The section on Teitelbaum to ref. (e)* is taken from Heywood, J. (2012). The response of higher and technological education to changing patterns of employment. *Proc. American Society for Engineering Education*, June 2012.

Teitelbaum (i)* clarified his thinking in a substantial treatise in 2014. He summarises his findings as follows:

- "First that the alarms about widespread shortages or shortfalls in the number of U.S. Scientists and engineers are quite inconsistent with the available evidence.

- second that the similar claims of the past were politically successful but resulted in a series of booms and busts that did harm to US science and engineering and made careers in these fields increasingly unattractive, and

- third that the clear signs of malaise in the U.S. science and engineering workforce are structural in origin and cannot be cured by simply providing additional funding. To the contrary recent efforts of this kind have proved to be destabilizing, and advocates should be careful what they wish for." [(i)*, p. 3].

(i)* Teitelbaum, M. S. (2014). *Falling Behind: Boom, Bust and the Global Race for Scientific Talent.* Princeton, NJ, Princeton University Press.

[2] The term "technologist," is used in British official documents instead of "engineer" but it mostly refers to "engineer." 57

[3] Advisory Council on Scientific policy. Committee on Scientific Manpower (1961). *The Long-Term Demand for Scientific Manpower.* London, HMSO, Cmnd 1490. 57

[4] Advisory Council on Scientific Policy. Committee on Scientific Manpower. *Scientific and Technological Manpower in Great Britain 1962.* London, HMSO, Cmnd 2146. 57

[5] *Loc. cit.* Ref. [1, (i)*]. 57

[6] Charette, R. N. (2013). The STEM crisis is a myth. *IEEE Spectrum*, pages 41–52, International, Spectrum.IEEE.ORG, September. 57, 71

[7] *Loc. cit.* Ref. [1, (i)*]. 58, 74

[8] *Loc. cit.* Ref. [6]. 58

[9] *Loc. cit.* Ref. [1, (i)*, p. 57]. 58

[10] *Loc. cit.* Ref. [1, (i)*, Ch. 5]. 58

[11] *Professional Manager*, 2011, winter issue, p. 12. Chartered Management Institute. 58

[12] See Becker, S. F. (2010). Why don't young people want to become engineers? Rational reasons for disappointing decisions. *European Journal of Engineering Education*, 35(4):349–366. 58, 74

[13] Schneiderman (2010). Economy and shortages affect European job outlook. The bigger high-tech companies in Europe are recruiting EE's. Talent is in short supply, especially to smaller firms looking for very specific skills. *IEEE Spectrum*, March. See also *First Bell*, 12:03:2010. U.S. facing shortage of nuclear scientists, engineers. 19:09:2007. Energy experts warn of worker shortage in Southeast U.S. 14:01:2008. Hewlett-Packard CEO warns of lack of U.S.-trained engineers. 05:03:2008. Aerospace defence industries brace for worker shortages. 30:07:2008. Engineer shortage threatens oil supply growth, says oil executive. 58

[14] Corinna Wu (2011). *ASEE Prism*, p. 40, November. 59

[15] *The Professional Manager*, Winter 2011. Chartered Management Institute. 59

[16] Cited by Zachary, G. P. (2011). Jobless innovation? *IEEE Spectrum*, p. 8, April. 59

[17] Zachary, G. P. (2011). Jobless innovation? *IEEE Spectrum*, p. 8, April. 59

[18] Wadhwa, V. (2011). Leading edge: Over the hill at 40. *ASEE Prism*, p. 32. 59

[19] *Loc. cit.* Ref. [6]. 60

[20] Irwin, N. (2017). To understand inequality consider the janitors at two big companies then and now. The Upshot September 3, 2017. *The New York Times*. The companies contrasted are Eastman Kodak and Apple. The article gives a very good idea of the different cultures. This writer worked in firms with cultures similar to Eastman Kodak in the 1950s. 60

[21] Larkin, C. and Corbett, S. (2017). Paper on the financing of alternative models of higher education. At an invitation seminar on "Higher Education and Technological Disruption: Purpose, Structure, and Financing." Dublin City University, October 9, 2017. 61

[22] Cheville, R. A. (2016). Linking capabilities to functioning adaptive narrative from role playing games for education. *Higher Education*, 71:805–818. 61

[23] Susskind, R. and Susskind, D. (2015). *The Future of the Professions. How Technology will Transform the Work of Human Experts*. Oxford, Oxford University Press. 61

[24] Bloomberg, August 30, 2017. These robots are using static electricity to make NIKE sneakers. Describes a robot developed to make the uppers of sneakers that has the effect of reducing the work force considerably. In a Nike making factory in China the average employment is about 1,300. In a plant designed for the robots in the U.S. it is 130. 61

[25] Patel, P. (2017). Millennials are concerned that automation will reduce their job prospects. *The Institute (The IEEE news Source)*, May 10, 2017. IEEE.org (http://ieee.org). Cites Deloitte Millennial Survey (https://www2.delloite.com/global/en/pages/about-deloitte/articles/millenialsurvey.html). See also automation (http://theinstitute.ieee.org/ieee-roundup/blogs/blog/will-automation-kill-or-create-jobs). 62

[26] *Loc. cit.* Ref. [57]. 62

[27] *Loc. cit.* Ref. [57]. 62

[28] Susskind, D. (2020). *A World without Work. Technology, Automation, and How We Should Respond*. UK, Allen Lane (Penguin Random House). 62

[29] *Ibid.* "First is the economic problem […] how to share prosperity in society where the traditional mechanism for doing so, paying people for the work they do, is less effective than in the past," […] Second, "the rise of Big Tech, since in the future, our lives are likely to become dominated by a small number of large technology companies. In the twentieth century, our main worry may have been the economic power of corporations but in the twenty-first that will be replaced by fears about their political power instead" […] Third, "It is often said that work is not simply a means to a wage, but a source of direction: if that is right, then a world with less work may be a world with less purpose as well." 62

[30] Autor, D. H. (2015). Why are there still so many jobs? The history and future of workplace automation. *Journal of Economic Perspectives*, 29(3):3–30. 63

[31] *Ibid.* 63

[32] Brynjolfsson, E. and McAfee, A. (2011). *Race Against the Machine*. Lexington, MA, Digital Frontier Press.

[33] Information Technology and the U.S. Workforce. Where are and where do we go from here? (2015). *A Report of the National Academies of Sciences, Engineering, Medicine*. Washington, DC, National Academies Press.

"Because most jobs involve multiple subtasks, and because technology typically targets specific tasks, one common impact of technology is to shift the distribution of tasks the human worker performs in a job (e.g., authors today spend less time proof reading for incorrect spelling enabling them to spend more time on the content of what they are writing). Technology also makes new tasks and new jobs possible transforming the nature of work in many, and ultimately most, industries" (p. 138).

[34] For example, Ridley, M. (2016). Let's stop being so paranoid about androids. Pessimists have always warned that automation will abolish everyone's job, yet it continues to improve our lives. *The Times*, November 21.

[35] *Loc. cit.* Ref. [62]. 63

[36] *Loc. cit.* Ref. [62]. 63

[37] Holzer, H. J. (2015). *Job Market Polarization and U.S. Worker Skills. A Tale of Two Middles*. Washington, DC, Brookings Institution, Economic Studies Working Paper, April 6. 63

[38] Karsten, J. and West, D. M. (2017). Can augmented reality bridge the manufacturing skills gap? Brookings, 21/08/2017. https://www.brookings.edu/blog/techtank/2017/08/10can-aygmented-reality-bridge-the-manufacturing-skills-gap/?utm_campaign=Center%20for%20Technolog 63

[39] OECD (2015). *Science, Technology, and Industry*. Scoreboard, 2015. Paris, OECD Publishing. http://dx.doi.org/10.1787/emploutlook-2015-en 63

[40] OECD. Policy brief on the future of Work. *Skills for a Digital World*. Paris, OECD. 63

[41] Matthews, M. (2011). Editorial, November 2011. *ASEE Prism*. 63

[42] Cited by Sparks, E. and Waits, M. J. (2011). *Degrees for What Jobs? Raising Expectations for Universities and Colleges in a Global Economy*, pages 20, 23, National Governors Association. 64

[43] *Ibid.* Minnesota measures: 2009. *Report on Higher Education*, pages 22, 23, 26, 27, Minnesota Office of Higher Education. 64

[44] 04:01:2010 *First Bell* programs touted as helping students prepare for middle skill level jobs. 64

[45] *Occupational Outlook Handbook*. U.S. Bureau of Labor Statistics, Fast growing Occupations. Also Most New Jobs. Published together December 17, 2015. 64

[46] *Loc. cit.* Ref. [12]. 64

[47] Goodwin, B. (2012). Research says/don't overlook middle skill jobs. *Educational Leader-ship*, 69(7) provides a brief review. 64, 74

[48] *Loc. cit.* Ref. [7]. 64

[49] Baker laments "snobbery" against technical colleges. *The Times*, January 30, 2017. But the colleges sponsored by Lord Baker are controversial. See Gove, M. My lesson from the latest schools scandal. *The Times*, February 10, 2017. 64

[50] Career Vision. The Ball Foundation https://careervision.org/opprtunities-abound-mioddle-skill-jobs. Accessed 14:08:2017. 65

[51] *Ibid.* 65

[52] Rosenbaum, J. (2001). *Beyond College for All: Career Paths for the Forgotten Half.* New York, Russell Sage Foundation. Cited by Goodwin Ref. [47]. 65

[53] Acemoglu, D. and Restrepo, P. (2017). Secular stagnation? The effect of aging on economic growth in the age of automation. *Working Paper 23077*. Cambridge, MA, National Bureau of Economic Research. http://www.nber.org/papers/w23077 65

[54] Acemoglu, D. and Restrepo, P. (2017). Robots and jobs: Evidence from U.S. labor markets. *Working Paper 23285*. Cambridge, MA, National Bureau of Economic Research. http://www.nber.org/papers/w23285 65

[55] Hanushek, E. A., Schwerdt, G., Woessmann, L., and Zhang, L. (2017). General education, vocational education, and labor market outcomes over the lifecycle. *Journal of Human Resources*, 52(1):48–87. 66

[56] The ideas of fluid (Gf) and crystallized (Gc) intelligence were introduced by Raymond Cattell. Gc is related to what a person has learned and Gf relates to a person's ability to respond to new and different situations. The model was developed by John Horn to include other dimensions including short term memory. Recently, the model has been combined with a hierarchical model due to J. B. Carroll. 66

[57] Jenkins, S. (1995). *Accountable to None. The Tory Nationalisation of Britain*, p. 119, London, Hamish Hamilton. 66, 72

[58] *Ibid.* 67, 75

[59] The optimistic view is that all will be well if workers are re-skilled but this has implications for the institutions that provide for re-learning. It is argued that the fear of robots is misplaced. In many areas the human has the advantage over the robot which will be

increased in some circumstances when the human works in collaboration with a robot(s). In any event changes take place in the very long term because there are so many difficulties to be overcome with machine learning. The pessimistic view argues that these assumptions are based on previous experience, and the assumption that the past acts as a guide to the future no longer holds [58]. The optimistic view is the result of convergent visioning [59]. Rapid changes in technology and the probable development of new materials are likely to speed up the rate of technological change [60]. 67, 75

[60] Kourdi, J. (2010). The future is not what it used to be. *The Professional Manager*, 19(2):27–28, Chartered Management Institute. 67, 75

[61] The terms are taken from J. Guilford's theory of creativity where divergent as contrasted with convergent thinking is often taken as the base for creative thinking. See Gregory, S. A. and Monk, J. D. (1972). Creativity: definitions and models in S. A. Gregory (Ed.) *Creativity in Innovation and Engineering*. Butterworths, London.

[62] Aligning technology and talent development. Recommendations from the APLU and NCMS expert educator team. *Report 1*, Summer 2017. Association of Public Land Grant Universities and National Center for Manufacturing Services in Association with Lightweight Innovation for Tomorrow (LIFE) Manufacturing Institute. 73

CHAPTER 5

Education in Service of Employment

5.1 GRADUATES FOR INDUSTRY

Since the 1950s, there has been a continuing flow of criticism of the formal system of tertiary education in the UK. It caused the Ministry of Education to develop a sub-system of technological education that would provide graduates specifically for industry. For a number of reasons this system failed. The ten colleges (Colleges of Advanced Technology-CATs) selected to undertake this work became universities, having demonstrated that the courses they implemented were of degree standard. One reason for their failure to deliver a curriculum that was different to those offered by the universities was the lack of an adequate operational philosophy that would have enabled the embryo ideas that existed to be developed into a full-bodied curriculum, for example, the idea of "integration," that is the linkage became academic study and the industrial training in cooperative learning too which the colleges were supposed to be committed, was not thought through. The possibilities of what could be achieved were not understood, nor was the role of experience in learning [1]. Similarly, there were no evidence-based studies of what engineers actually did in industry to inform the curriculum. Although the system's policy was constructed around the beliefs of a few distinguished persons' views about engineering education it depended on what the academics in these institutions could and/or wanted to do.

By the time *The Taxonomy of Educational Objectives* was published in the UK (1964), the CATs knew they were going to become universities [2]. *The Taxonomy* might have raised the level of discussion to the extent that an educational framework may have emerged that might have led to the assessment of, and thus learning for, higher-order thinking skills. This view is supported by the successful development of Engineering Science at the Advanced level of the General Certificate Education, an examination used for entry to university [3], although its designers found that the categories of the taxonomy did not completely meet their needs, particularly with respect to creativity.

An attempt in the early 1970s to task analyse the work done by engineers in a firm in the aircraft industry for the purpose of developing a taxonomy of training objectives showed that any such taxonomy should include categories of communication, diagnosis, and management [4]. In the 40 or so years since that study, industrialists have continually complained about the poor communication skills of graduates. While universities have taken steps to develop communication skills in engineering courses, James Trevelyan, as a result of analysing the practice of engineers, concluded that these courses did not meet the needs of industry. He found the key work done by engineers was "technical liaison" in which a form of communication is a priority skill [5].

An evaluation among managers and supervisors of the British Steel Corporation who had attended in-company training courses showed that the most important need was for communication skills.

Bill Humble, a senior training officer with the company, who with this writer undertook that study, also analysed management and supervisory tasks using the categories of both the affective and cognitive domains of *The Taxonomy*. They showed the importance of skills in the affective domain [6]. (In this text the affective domain is taken to embrace values, beliefs, and behaviour, and encompass those qualities that are commonly thought to belong to the rather loose concepts of social and emotional intelligence [7].) Some persons may think of the combination as representing "personality." That said, while there is increasing recognition of this domain we have not reached the stage where, with Macmurray we can assent to the view, "that the intellectual mode of reflection is a derivative from the emotional, and is contained within it" [8, p. 200].

That skills in both domains are required by managers and supervisors is illustrated by another study undertaken by Humble who observed managers and workers in situations that were to some extent confrontational. His partially developed categories are shown in Exhibit 5.1. Notice, that he includes a category of "adaptability" and defines what it means. But it is self-evidently incomplete and will be explored in the next chapter since demands that persons are adaptable, whatever that might mean, continue to be made.

Humble defines a category of "control" in terms of the management of people. Currently, there is surely need for a category of "control," not only of people, but of technology and ourselves. For the most part, individuals adapt to what is available. People have readily taken to devices that improve family life, e.g., washing machines, dish washers. They change their mobile (cell) phones with every update, and they accept much of what is forced them as, for example, changes in methods of banking. Swedes are now accepting hand implants of micro-chips that enable them to make financial transactions, and so on.

The most worrying thing is that the public seems to have little concern for such matters. For example, Jaron Lanier seems not to have been read, or if he has, we have not been persuaded that the IT companies should pay us for the information they have about us; yet, it is quite obviously one way of controlling big tech which has become big tech because of the information it has been given [9].

But, within the last year or so, we have learnt that we have to be much more critical of what is placed before us, and this requires higher-order thinking.

It is not surprising that Humble should have included a category of "relationships," since he had to deal with conflict situations. Recently, this category has been expressed as the ability to work in teams.

The ability to adapt involves:

The ability to perceive

1. the organizational structure and formal, informal relationships, value systems and languages, and therefore its needs;
2. knowledge of the technical, human, and financial aspects of the system or situation;
3. the different thought processes involved in the solution of human or technological problems; and
4. our own self and attitudes.

The ability to control involves:

1. knowledge of
   (i)   how the skills of those who have to be controlled should be used;
   (ii) his or her requirements in relation to needs for communication, competence, and excellence;
   (iii) what people ought to be doing;
   (iv) whether or not they are doing it effectively; and
   (v)   how to create a climate in which jobs will be done effectively,
2. the ability to make things happen,
3. the ability to be able to discriminate between relevant and irrelevant information, etc.

The ability to relate with people involves:

1. knowledge of rights, responsibilities, and obligations;
2. knowledge of ways of thinking (determinants of attitudes and values) of people in all parts of the organization;
3. ability to understand when action in the key environment is right and acceptable in those circumstances (i.e., to understand the effect of his or her behavior on a situation);
4. ability to be able to predict the effects of his or her behavior and that of others on a situation; and
5. ability to create the feeling that the job is important.

Exhibit 5.1: Humble's partial derivation of a taxonomy of industrial objectives from a typical works (steel) situation in which managers and workmen were in some degree of confrontation. Circa 1966.

5.2   SHAPING THE ENVIRONMENT AND STERNBERG'S THEORY OF INTELLIGENCE

Humble's second category of control required that we should be able to shape (control) our environment. Nearly 20 years later Robert Sternberg defined intelligence as a "mental activity directed towards purposive adaptation to, selection and shaping of, real world environments relevant to one's life" [10]. This is very much about individual's controlling and directing them-

1. **Practical problem solving ability:** Reasons logically and well identifies connections among ideas, sees all aspects of a problem, keeps an open mind, responds to other's ideas, sizes up situations well, gets to the heart of the problem, interprets information accurately, makes good decisions, goes to original sources of basic information, poses problems in an optimal way, is a good source of ideas, perceives implied assumptions and conclusions, listens to all sides of an argument, and deals with problems resourcefully.

2. **Verbal ability:** Speaks clearly and articulately, is verbally fluent, converses well, is knowledgeable about a particular field, studies hard, reads with high comprehension, reads widely, deals effectively with people, writes without difficulty, sets times aside for reading, displays a good vocabulary, accepts norms, and tries new things.

3. **Social competence:** Accepts others for what they are, admits mistakes, displays interest in the world at large, is on time for appointments, has social conscience, thinks before speaking and doing, displays curiosity, does not make snap judgements, assesses well the relevance of information to a problem at hand, is sensitive to other people's needs and desires, is frank and honest with self and others, and displays interest in the immediate environment.

Exhibit 5.2: Abilities which contribute to intelligence. Obtained from questions about the nature of intelligence, academic intelligence, and unintelligence put to experts in research on intelligence and lay persons by R. H. Sternberg and his colleagues. Among the findings was the fact that research workers considered motivation to be an important function of intelligence whereas lay persons stressed interpersonal competence in a social context. In R. J. Sternberg (1985) *Beyond IQ. A Triarchic View of Intelligence*, Cambridge University Press.

selves. Given the little *Oxford Dictionary's* definition of management as "direction and control" it is about self-management.

Comparing Humble's categories to the categories that Sternberg derived from a study of the definitions that lay people and experts gave of intelligence shown in Exhibit 5.2 [11] suggests many similarities between them. This is equally true of the many lists of objectives (abilities) drawn up to indicate skills that their authors believe need to be learnt. This correspondence may be indicative of the idea put forward by Bernard Lonergan that there is a method of human understanding underlying all common sense and scientific knowing and that practical decisions are an extension of that method [12].

In the decades that followed Sternberg and his colleagues have pursued the concept of "practical intelligence" which they define as "intelligence used in real world contexts" [13]. One of the examples they give is that "you might want to try and change the requirements of a new job to suit yourself." They would regard the process involved in achieving this goal as adaptive behaviour. They believe that "practical intelligence" can be developed, and as an aid, they designed an adaptive behaviour checklist, and suggested four exercises for the development of practical intelligence. They write "practical problems unlike academic ones, often have no single right or wrong answer. Indeed, the mistake people often make is looking for the certainty that just does not exist in practical situations. Practical situations of any importance almost always involve some elements of risk, uncertainty, and ambiguity." There is no better example of that than the problems that politicians and the public were asked to solve as a result of the Covid-19 pandemic.

Some engineering educators design "wicked" problems with a view to replicating real world problems but industrialists continue to complain that graduates are not prepared for such scenarios.

Practical intelligence is an intelligence of action, "I do." This is what industrialists want graduates to do. It is at the heart of the difficulties that arise between academics and industrialists It is the "I think" vs. the "I do." Macmurray writes, "The effect of transferring the centre of reference to action, and at the same time its sufficient justification, is that man recovers his body and becomes personal, when he is conceived as agent, all his activities, including his reflective activities, fall naturally into place in a functional unity. Even his emotions instead of disturbances to the placidity of thought, take their place as necessary motives which sustain his activities, including his activity of thinking" [14, p 12]. Macmurray invites us to think of the interdependence of the affective and cognitive as the domain of the personal.

Examination of the complaints of industrialists since Humble derived his objectives show that, while they do not use either the term affective or personal, the skills they seek relate as much to the personal as they do to the purely cognitive, if not, more so. It is by no means clear that academics understand this to be the case. The related terminology been subject to change. These skills have been called "soft" by engineering educators but they now prefer the term "professional." In the UK they were first called "personal transferable skills" (see below).

As will be seen, industrialists in the UK have made similar complaints about graduates in general, and governments' and quasi-official agencies have responded.

5.3   RESPONSES BY POLICY MAKERS TO INDUSTRIAL COMPLAINTS

In 1989 in response to complaints by industrialists about the performance of new graduates, the UK's Employment Department (equivalent U.S. Department of Labor) initiated a programme designed to encourage all universities to develop the skills that industry claimed are required. It was called the Enterprise in Higher Education Initiative (EHEI), and the skills it developed were called the skills of enterprise learning. As part of its work the programme supported a research and development unit at the University of Sheffield that sought to clarify what these skills were. It achieved this goal by analysing hundreds of adverts for graduates in the national press and came up with the model shown in Exhibit 5.3 [15]. It will be seen that it has nothing to do with knowledge content but is related to the cognitive and affective behaviours that graduates should bring to their work. They called them "personal transferable skills."

TRANSFERABLE PERSONAL SKILLS
A developmental model

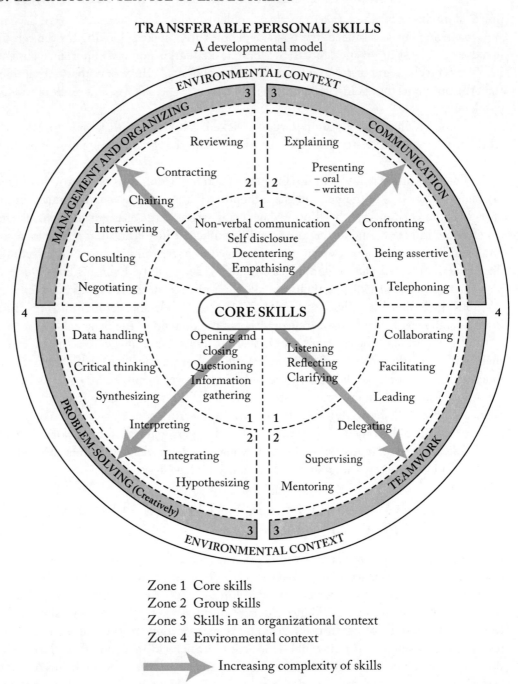

Zone 1 Core skills
Zone 2 Group skills
Zone 3 Skills in an organizational context
Zone 4 Environmental context

Increasing complexity of skills

Exhibit 5.3: Personal Skills Chart developed by the Personal Skills Unit of Sheffield University.

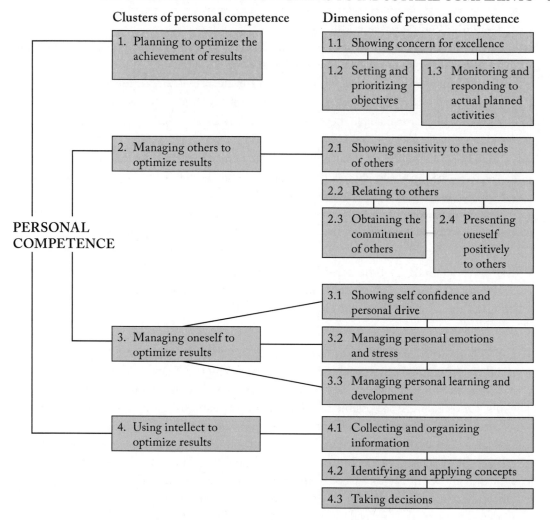

Exhibit 5.4: The Personal Competence Model developed by the Management Charter Initiative funded by the Employment Department, *Personal Competence Project*, Summary Report 1990 Standards Methodology Branch, Employment Department, Sheffield.

Universities in the UK were financially supported by the Employment Department for five years to develop these skills within all subjects.

The Employment Department wanted universities to assess these skills and the committee that was convened to develop assessment drew up the list of areas of learning that are important for equipping students for their working lives (Exhibit 5.5). The influence of the Sheffield study will be apparent as will the similarities with the views of intelligence expressed by lay persons

Cognitive knowledge and skills
- **Knowledge:** Key concepts of enterprise learning (accounting, economics, organizational behavior, inter- and intra-personal behavior).
- **Skills:** The ability to handle information, evaluate evidence, think critically, think systemically (in terms of systems), solve problems, argue rationally, and think creatively.

Social skills: as for example the ability to communicate and work with others in a variety of roles both as leader and team leader.

Managing one's self: as for example, to be able to take initiative, act independently, take reasoned risks, want to achieve, be willing to change, be able adapt, know one's self and one's values, and to be able to assess one's actions.

Learning to learn: to understand how one learns and solves problems in different contexts and be able to apply the styles learned appropriately to the solution of problems.

Exhibit 5.5: The four broad areas of learning together with the elements they comprise that are important for equipping students for their working lives, as defined by the REAL working group of the UK Employment Department, 1991.

and experts in Sternberg's study (see Exhibit 5.2). The Employment Department also funded the development of a personal competence model by the Management Charter Initiative. The model is shown in Exhibit 5.4. It clearly depends on the development of personal transferable skills. Once again, it is possible see elements of the categories developed by Humble.

The U.S. Department of Labour was also concerned with secondary education as a preparation for the world. It considered that a high school diploma was worthless and proposed in 1992 a high school curriculum (SCANS curriculum) suited to the world of work [16]. Exhibit 5.6 shows the work place competencies and foundation skills around which the SCANS curriculum should be developed. It should be noted that systems (thinking) is one of the categories. Of particular interest are the examples of how these skills could be integrated into the curriculum given in the report. Some schools did offer a SCANS type curriculum but it was not an idea that took off. Might that be because it came from a Department of Labour? In the UK the Employment Department failed to maintain significant initiatives that it made in schools (e.g., The Technical Vocation Education Initiative (TVEI)).

The personal transferable skills have many similarities with those listed by the U.S. State of Minnesota Office of Higher Education. A re-arranged and simplified list of the top skills required by Minnesota employers is shown in Exhibit 5.7. It can be inferred from this list that the affective domain, and personal qualities are likely to be as important as the cognitive. They are not skills that separate the academic from the vocational, but skills that any educated person would require, and that liberal educators profess to develop. The Employment Department believed that these skills could be developed within the undergraduate subject a person studied provided

Workplace Competencies	Effectively Workers Can Productively Use
1. Resources	They know how to allocate, time, money, materials, space, and staff.
2. Interpersonal skills	They can work in teams, teach others, serve customers, lead, negotiate, and work well with people from culturally diverse backgrounds.
3. Information	They can acquire and evaluate data, organize and maintain files, interpret and communicate, and use computers to process information.
4. Systems	They understand social, technological, and organizational systems, they can monitor and correct performance, and they can design or improve a system.
5. Technology	They can select equipment and tools, apply technology to specific tasks, and maintain and troubleshoot equipment.
Foundation Skills	**Competent Workers in a High Performance Workplace Need**
1. Basic skills	Reading, writing, arithmetic, and mathematics. Speaking and listening.
2. Thinking skills	The ability to learn, reason, think creatively, make decisions, and solve problems.
3. Personal qualities	Individual responsibility, self-esteem and self-management, sociability, and integrity.

Exhibit 5.6: The SCANS competencies (SCANS = Secretary's Commission on Achieving Necessary Skills, Competencies and Foundation Skills for Workplace Know-How). *Learning a Living: A Blueprint for High Performance.* Washington, DC, U.S. Department of Labour.

that learning was designed for that purpose. The Sheffield Unit showed how this might be done. But it was also argued that: (a) some special provision should be made for separate study in what might be called organisational behaviour, and behaviour in organisations; and (b) that university programmes should be designed to meet the developmental needs of students, a proposal that had the theories of Perry [17] and King and Kitchener in mind [18].

In the U.S. in a challenging report for the National Governors Association in 2011, Erin Sparks and Mary Jo Waits who described the Minnesota Study reported above also described similar work in other States. They argued that the States should take much more notice of the comments of employers than they had in the past. More dramatically, they proposed to the National Governors Association that the Governors' should make a radical change in direction, and redirect their support away from traditional four-year programmes of general education

Attributes (most frequent ratings of "very important" by employers)
   Professionalism (punctuality, time management, attitude)
   Self-direction, ability to take initiative
   Adaptability, willingness to learn
   Professional ethics, integrity
   Verbal communication skills

Most frequent ratings of "not at all" or "not very important"—last five items
   Advanced mathematical reasoning (linear algebra, statistics, calculus)
   Technical communications
   Fluency in a language other than English
   Knowledge of specific computer applications required for the job
   Application of knowledge from a particular field of study

Other
   Capability
   Creativity
   Ability to work in a culturally diverse environment
   Ability to work in teams
   Written communication skills
   Basic mathematical reasoning (arithmetic, basic algebra)
   Critical thinking and analysis
   Problem solving, application theory
   General computer skills (word processing, spread sheets)
   Knowledge of technology equipment required for job

Exhibit 5.7: A re-arranged summary of the 20 top skill needs of Minnesota Employers. Minnesota Office of Higher Education. Minnesota Measure. 2009 Report on Higher Education Performance. Cited in full in Sparks, E. and Waits, M. J. (2011). *Degrees for What Jobs? Raising Expectations for Universities and Colleges in the Global Economy*, p. 27, Washington, DC, National Governors Association.

towards the provision of courses that would *"better prepare students for high paying, high demand jobs"* [19].

   This substantial paradigm shift briefly reignited the debate between two opposing philosophies about the relative merits of liberal/general education on the one hand and, on the other hand, job-oriented (vocational) education [20]. The report should have provided an opportunity to open a fundamental debate about the aims of education in a technological age that might have been taken up internationally. That did not happen, and decisions worldwide continue to be made that are based on the prevailing utilitarian philosophy. (In particular, they ignore research in pedagogy and curriculum.) In essence, current discussions about engineering education revolve around this debate. Its resolution in the light of technological change is also its relevance to higher education *per se*.

5.4 THE DEVELOPMENT OF PROFESSIONAL COMPETENCE: RECONCILING TWO PHILOSOPHIES

Introduction

A major impact of the utilitarian philosophy of engineering education caused by the relentless search for profits by large organisations, is the demand that graduates should be prepared to work immediately in industry, that is, to do a specific job. The question arises, therefore, as to whether traditional courses can or should prepare students immediately to slot into a job in industry.

Previous analyses have suggested that there are limits to what college education can achieve [21]. For example, one paper suggested that even in project-based team activities students could acquire or reinforce attitudes already acquired that are inimical to industry's requirements [22]. It led to the argument that selectors should recruit students who already had experience of industry which draws attention to the value of internships and sandwich (cooperative courses).

Clearly, depending on their structure, sandwich (cooperative) courses would have a much better chance of achieving that goal. But, as this section will demonstrate, the success of an immediate placement will depend on the view that the firm takes of competency for the evidence suggests that it is necessary to understand that competency relates to the job, and is learnt as a function of the job environment. That environment may impede or enhance that learning. In some environments the acquisition of engineering competency may take years, because just as school is not the end of personal and cognitive development, neither is college.

Nevertheless, complaints about the quality of graduates continue to be made in spite of substantial attempts by some institutions to better prepare graduates for work [23]. Attempts to reform the curriculum make up for what are perceived to be deficits in the curriculum, and in so-doing develop professional competence [24]. This has caused much concern among engineering educators who believe that these additions create deficits in the curriculum they have to teach. All of this has taken place without any adequate understanding of the philosophical and socio-psychological issues involved. Nevertheless, a more profound debate between academia and industry could take place if the two sides reflected on the principles that drive their thinking. This section is intended to outline the bases for such a debate.

Outcomes and "Inside" Competence

Fundamental to the outcomes approach which currently dominates curriculum policy, is a view of learning that is deeply embedded in the Western Psyche. Catherine Griffin writes, "Machine metaphors, common in western conceptions of the mind and thinking, also define what is involved in being a competent person. In many European and American cultural contexts, the person is represented and realised as a separate autonomous entity, that is, an individual. Individual actions result from attributes of the person that are activated and then cause behaviour. Individual actions result from the attributes of the person that are activated and then, cause behaviour. Accordingly 'competence' is located 'in' the individual, 'in' the mind, 'in' the brain. This view of competence is active, as in the machine metaphor, it cranks, works, churns, and then out comes the solution to the problem" [27]. Many academics and industrialists believe this to be the case. Therefore, change your syllabus and assess performance of the syllabus, and students will be prepared for industry. Alas, no.

Those industrialists who accept this "inside" view of competence neglect to notice that while it is Cartesian in demanding that students think, industry in contrast also wants its personnel to act. Industry is aligned to the view expressed by Macmurray that all our theoretical activities derive from our need to solve practical problems [28]. It is aligned with an "outside" view of competence which holds that it (competence) is developed through relationships with others in their social (work) environments. Thus, competency is context dependent and assessable. In this context, relationships are as important as they are in college and require a person to be as competent in the "soft" skills (affective domain) as they are intellectually (cognitive domain).

The Problem of Integration

Understanding that in these exchanges the self is agent provides a basis for a theory of integration. For example, an initial internship would provide experience that leaves questions unanswered or explanations not fully understood that would be illuminated in college, but it would be preceded by some an induction that provides the student with skill in connoisseurship (see Section 7.6). This places an obligation on colleges and industrial organisations that the desired integration happens. It also follows, given that the self-is-agent, that an obligation is placed on the company to work with the student to achieve mutually agreed goals.

An investigation by Korte [29] of the experience of the workplace of young graduates shows how this might work. A student told him that he *"wish(ed) someone had told him how to play the political game here."* He needed to know about the informal organisation of the company. "Informal Organisation" is one of the earliest ideas of organisational theory that is not much considered today, but from the perspective of a person beginning a new job it is a very useful concept. Understanding the informal networks within an organisation leads to an understanding of how power is distributed and influences the movement of the organisation towards its goals. It has its equivalent in the hidden curriculum in school which is an important reminder that many factors other than the cognitive contribute to learning and success at work. Korte (personal communication) would argue that it is possible to use the students' experiences of the hidden curriculum to sensitise them to what they are likely to experience in industry and how to interpret it.

If students receive a prior induction course that indicates what they should look out for in an internship then much more can be made of management studies during the academic course. More can be made of the internship if the students are also introduced to the notion of reflection. It may be argued, that if the curriculum follows the Kolb cycle of experiential learning, that only a period in industry is likely to provide the experience that is necessary for reflection, provided the academic course is designed to draw out that capacity (*educare*). This principle may be extended to cooperative courses. If an industrial period is arranged to follow an academic period then it becomes possible to provide the students with insights that will help them understand more easily how the organisation functions and how engineering theories play out in practice. Such insights cannot come by osmosis from working in teams in college.

"Outside" Competence

This emphasis on the experiential is in contrast to the philosophy of teaching that is promoted by the "inside" model of competence. Catherine Griffin wrote, "When competence is viewed as a dynamic process it is defined as the ongoing acquisition and consolidation of a set of skills needed for performance where intrinsic motivation is the driver, Inside (innate individual properties) models of competence describe individuals as 'having' varying amounts of competence whereas 'outside' models of competence focus on external contextual, social, cultural, and historical factors which either hinder or contribute to the development of an individual's 'competence' [30]. At this point social-psychology takes over and finds support for the 'outside' model of competency."

Sandberg, a Swedish investigator, showed quite clearly that a particular competence required by a firm in the automobile industry was context dependent [31]. It was representative of the "outside" model of competence. He found that the particular competence he studied was possessed at different levels of skill among the engineers performing the task with which it was associated. The task had to be done before training needs could be identified and decisions made about the added-value of training, if any. There is always learning on the job and some of this is continuous which is why firms are learning organisations that can organise themselves so that they enhance or impede learning. Those that impede learning will inevitably die (see Chapter 6 [32]). College programmes can prepare students for organisational learning but they cannot provide the situations in which engineers will find themselves for the competences required will be particular to the organisation.

Blandin [33] gives a different meaning to "levels" to Sandberg. He writes "[...] it can be said that 'competency' is an ingredient of 'competence.' But competence is more than the sum of its ingredients. In fact when compiling the literature, it appears that 'competence' develops and is recognised at three levels: at the level of the individual, at the level of the group in which the individual works, and at the level of the organisation in which he/she works. At the level of the individual, competence has a cognitive dimension [...] the result is a demonstrated level of proficiency and a feeling of self-efficacy." The importance of self-efficacy is to be found in a number of studies in the engineering education literature.

Blandin's model of "competency" takes into account these three dimensions. Hence, the knowledge, skills and procedures relate to and are affected by the context and indicators of competence. Blandin and his colleagues found that the cognitive dimension of competence had five main competency indicators. One "acting as engineer in an organisation" was found to be more important than the others, and was thus the core competency. It "develops only within the company and cannot exist without long experience within a company." Within the company the students who were technicians seeking to graduate as engineers developed competencies that were specific to their job. The writer takes from this study that the interaction between periods of academic study and industrial work help students to acquire professional competence that is not available to courses of the traditional kind that have no industrial contact.

It is concluded that the development of professional competence is as much the responsibility of industry as it is of educational establishments, and this requires a joint approach to assessment in which a portfolio is likely to have a major role.

5.5   CONCLUSION

A major impact of the utilitarian philosophy of engineering education has arisen from the relentless search for profits by large organisations. It is that graduates should be prepared to work immediately in industry. Complaints about the quality of graduates continue in spite of substantial attempts by some institutions to better prepare graduates for work these complaints continue. Within the framework of a utilitarian philosophy the matching of the outputs of education to the inputs (needs) of industry might be regarded as its moral purpose.

Many lists of outcomes have been published. Some derive from published documents like *The Taxonomy of Educational Objectives*, others from alternative theories of intelligence and still more from observations of what engineers, supervisors and managers do in their jobs. There is much agreement between the different lists. When they are compared with the list of qualities that contribute to intelligence obtained from experts and laypeople by R. J. Sternberg and his colleagues it seems that all of the inventories are seeking from individuals "intelligent behaviour," and that is far from measures of academic intelligence. Make a further comparison with the goals of liberal education and it seems there is a demand for people who have had such an education, surprising as this may seem [34, pages 83–86].

Both directly and indirectly these two lists draw attention to the need for employees to be adaptable and flexible by which is meant the ability to learn a new job when the one they are in becomes redundant. What was a cliché has become a reality for employees, but it brings with it responsibilities for employers. Continuing personal (professional) development will have to become a *sine qua non*, and a responsibility of both employee and employer, so that employees have the skills necessary for transfer to a wide range of occupations. For this they will require a wider and more unified form of knowledge than in the past. A proposal for an alternative model of higher and technological education set within a philosophical framework follows in the remaining chapters.

Just as in medicine the concept of competency has become the subject of much discussion in engineering particularly as it relates to accreditation and quality assurance [35].

The outcome model developed by ABET did not consider what it was to perform competently or whether competence could be taught in spite of a substantial literature in medical education on competency-based performance and assessment. Had it done so it would have found research that supported an "outside" philosophy of competency, rather than the "inside" model that dominates much thinking in engineering education—namely, that it is a quality innate to the individual, a consequence of which is that can be taught.

The "outside" model shows that competence is context dependent and that each workplace requires competencies that can only be acquired in the workplace. This is not to say that universities have no role to play in preparing students for the workplace. At the very minimum they can be taught how to observe what is happening in an organisation and establish what the organisation needs from them. Such activities focus on personal skills, the human side of engineering. More ambitious programmes would be designed cooperatively. But such activities do not release industry from its obligation to continue the development of the student.

A major reason for lack of response is that education is not governed by the search for truth but by cherished beliefs that are deeply embedded in particular culture (see Section 4.7). This is illustrated by Sir Charles Carter in a critical review of the recommendations of the 1963 Robbins Report on Higher Education in the UK written 20 years after the event [36]. He made the point they might have considered developing a model that contained institutions like the American community colleges. But that would have been too much a jump to make.

The idea that education like any other area of knowledge is open to examination is almost *verboten*. In these circumstances changing an education system requires substantial disruption. It is proposed here that rapid changes in technology and AI will cause that disruption. Irrespective of whether it is in the short or long run it behoves the higher education community to consider the possibilities in order for it to react effectively to such change.

Evaluation of one major attempt to change a system in the 1960s shows that any proposal for change has to be plausible to those who have to make the change which means that there has to be prior knowledge of the proposal that has to have been debated [37]. As Noddings' pointed out those who seek change must do so within an adequate philosophical framework [38].

5.6 NOTES AND REFERENCES

[1] Heywood, J. (1969). An evaluation of certain post-war developments in higher technological education. Lancaster, Thesis, University of Lancaster Library. Two Volumes. 77

[2] Bloom, B. et al. (Eds.) (1956). *The Taxonomy of Educational Objectives. Vol. I. Cognitive Domain*. New York, David Mackay. 77

[3] Joint Matriculation Board (1973). *Notes for the Guidance of Schools. GCE Engineering Science (Advanced)*. Manchester, Joint Matriculation Board. 77

[4] Youngman, M. B., Oxtoby, R., Monk, J. D., and Heywood, J. (1978). *Analysing Jobs*. Aldershot, Gower Press. 77

[5] Trevelyan, J. (2014). *The Making of the Expert Engineer*. London CRC Press/Taylor & Francis. 78

[6] Humble, W. cited by Heywood, J. (1970). Qualities and their assessment in the education of technologists. *International Journal of Mechanical Engineering Education*, 9:5–10. 78

[7] Krathwohl, D., Bloom, B., and Mersia, B. B. (1964). *Taxonomy of Educational Objectives. Vol. 2. Affective Domain*. London, Longmans Green. See also Kaplan, L. (1978). *Developing Objectives in the Affective Domain*. San Diego, CA, Collegiate Publishing. 29, 78

The categories of the Taxonomy's affective domain are, receiving, responding, valuing, organisation, characterisation by a value or value complex.

[8] Macmurray, J. (1958). *The Self as Agent*. London, Faber and Faber. 78, 94

[9] Lanier, J. (2013). *Who Owns the Future?* London, Allen Lane, Penguin. 78

[10] Sternberg, R. J. (1985). *Beyond IQ. A Triarchic Theory of Intelligence*. New York, Cambridge University Press. 79

[11] *Ibid.* 80

[12] Frezza, S. T. and Nordquest, D. A. (2019). Engineering insight: The philosophy of Bernard Lonergan applied to engineering in Korte, R., Mina, M., Frezza, S. T., and Nordquest, D. A. *Engineering Epistemology: Pragmatism and the Generalised Empirical Method*. San Rafael, CA, Morgan & Claypool. 80

[13] Sternberg, R. J., Kaufman, J. C., and Grigorenko, E. L. (2008). *Applied Intelligence*. New York, Cambridge University Press. 80

[14] Macmurray, J. (1962). *Persons in Relation*. London, Faber and Faber. 81

[15] Green, S. (1990). *Analysis of Personal Transferable Skills Requested by Employers in Graduate Recruitment Advertisements*. Mimeo, Sheffield, University of Sheffield. Personal Transferable Skills Unit. 81

[16] SCANS (1992). *Learning a Living. A Blueprint for High Performance. A SCANS Report for America*, 2000, Washington, DC, Department of Labor. 84

[17] Perry, W. G. (1970). *Forms of Intellectual and Ethical Development in the College Years*. New York, Holt, Reinhart and Winston. 85

[18] King, P. M and Kitchener, K. S. (1994). *Developing Reflective Judgement*. San Francisco, Jossey-Bass. 85

[19] Sparks, E. and Waits, M. J. (2011). *Degrees for What Jobs? Raising Expectations for Universities and Colleges in a Global Economy*. Washington, National Governors Association. 86

[20] The term "vocational" is used in a number of ways throughout the world. It is used to relate to particular careers that are said to be a "calling," e.g., nursing, the priesthood. It is also used as the opposite of academic to describe courses that rely on specialist techniques, e.g., technicians, clerical officers. It is used in the second sense here. 86

[21] Heywood, J. (2016). *The Assessment of Learning in Engineering Education. Practice and Policy*. Hoboken, NJ, IEEE Press/Wiley. 87

[22] Leonardi, S. M., Jackson, M. H., and Diwan, A. (2009). The enactment-externalization dialectic rationalization and the persistence of counterproductive technology design practices in student engineering. *Academy of Management Journal*, 2(2):400–420. 87

[23] For example Barry, et al. [24] completely redesigned a traditional lecture course in water and waste water treatment so that the students would learn about professional competences in addition to content. They called the new approach "challenged-based," which is a project-based approach to problem-based learning in that "*instruction begins with the presentation of a term project as a culminating event for the students' learning.*" Formative assessment involved the students helping with the design of the rubric which, it was argued, would help them to think about their own criteria for success. The final assessment was an oral defence of their work by the students individually with the instructors. 87

A consortium has developed an Integrated Design Engineering Assessment and Learning System (IDEALS) which targets professional development. The professional skills assessment instruments have been piloted in six different universities across the United States. Students are reported as perceiving the instruments add-value to their programme [25].

Relevant to these development are the development of taxonomies for the Teaching Engineering Practice, and Engineering Design Tasks [26].

[24] Barry, B. E., Brophy, S. P., Oakes, W. C., Banks, K. M., and Sharville, S. E. (2008). Developing professional competencies through challenge to project experiences. *International Journal of Engineering Education*, 24(6):1148–1162. 87, 93

[25] McCormack, J. et al. (2011). Assessing professional skill in capstone design course. *International Journal of Engineering Education*, 27(6):1308–1323. 93

[26] McCahan, S. and Romkey, L. (2014). Beyond Bloom's. A taxonomy for teaching engineering practice. *International Journal of Engineering Education*, 30(5):1176–1189. 93

[27] Griffin, C. (2012). A longitudinal study of portfolio assessment to assess competence of undergraduate nurses. Doctoral dissertation. Dublin, University of Dublin. 88

[28] *Loc. cit.* Ref. [8]. 88

[29] Korte, R. F. (2009). How newcomers learn the social norms of an organization: A case study of the socialization of newly hired engineers. *Human Resource Development Quarterly*, 20(3):285–306. 88

[30] *Loc. cit.* Ref. [30]. 89, 94

[31] Sandberg, J. (2000). Understanding human competence at work. An interpretive approach. *Academy of Management Journal*, 43(3):9–25. 89

[32] Heywood, J. (1989). *Learning, Adaptability, and Change. The Challenge for Education and Industry.* London, Paul Chapman/Sage. 89

[33] Blandin, B. (2011). The competence of an engineer and how it is built through an apprenticeship program: A tentative model. *International Journal of Engineering Education*, 28(1):57–71. 89

[34] Heywood, J. (2000). *Assessment in Higher Education. Student Learning, Teaching, Programmes, and Institutions.* London, Jessica Kinglsey. 90

[35] Pears, A., Daniels, M., Nyfen, A., and McDermott, R. (2019). When is quality assurance a constructive force in engineering education. In the Press for *ASEE/IEEE Proc. Frontiers in Education Conference*, Cincinnati, OH, October. 91

[36] Carter, Sir Charles (1983). Great Expectations. Robbins III Charles Carter recalls the conventional wisdom of the report, *The Times Higher Education Supplement*, p. 11, November 4, 1983. 91

[37] Heywood, J. (2006). Factors in the adoption of change: Identity, plausibility, and power in promoting educational change. *ASEE/IEEE Proc. Frontiers in Education Conference*, T1B:9–14, 2006. 91

[38] Noddings, N. (1995). *Philosophy of Education.* Boulder, CO, Westview Press. 91

See also Noddings, N. (2009). The aims of education in Flinders, D. J. and Thornton, S. J. *The Curriculum Studies Reader*, p. 425, 3rd ed., New York, Routledge.

CHAPTER 6

Adaptability, Transfer of Learning, and Liberal Education

6.1   ADAPTABILITY AND THE TRANSFER OF LEARNING

It is by no means clear what those who require persons to be adaptable mean by adaptability. Currently, it often seems to mean adaptation to a new technology that an organisation wishes to introduce. This has been a problem with automation since its beginning. Early studies of the impact of automation on the workforce and its adaptability spawned the concept of the socio-technical system and socio-technical systems theory [1]. While flexibility is clearly related to adaptability it can refer to those situations where individuals capable of doing other work are reluctant to do that work for one reason or another. The UK is sold to industrialists on the grounds that its labour market is more flexible than those in other countries, for example, France.

Of particular concern is the adaptation that individuals have to make when they transfer jobs. But what is adaptation? Looked at from the perspective change, the factors that impede innovation in an organisation, or prevent a society from adapting to a new circumstance, are the same factors that inhibit individuals from learning. From the perspective of change, individuals and organisations may be regarded as learners or learning systems. Neither individuals nor organisations can adapt if they cannot learn. This why over-reliance on experience and the tacit knowledge it creates can be inhibitive of development and innovation [2]. Thus, a key function of those who govern, lead, or manage is to create an environment conducive to learning. Since, according to Sternberg, intelligence is the ability to select and to shape the real world in which we live (see Chapter 5 and [3]), we too, have a responsibility to help those whom we allow to lead and manage, to create that environment [4].

Given that individuals are expected to continue learning it is surprising that so little attention is paid in higher education to the exploration of learning by students.

If learning and decision making are held to be the same thing, that is, goal seeking behaviours, then the activity of learning is an adaptive response. It depends on what is commonly called the "transfer of learning," and that may not be accomplished without difficulty. Students have to understand what transfer is in order to develop skill in its application.

Academic subjects in engineering and science are structured so that beginning from a study of fundamentals the learner develops existing skills and knowledge, and also acquires new knowledge and skills in order to be able to deal with any problems they may face when they have completed their hierarchically ordered studies. Mostly, they accomplish this without any formal study of learning. The ability to solve unforeseen problems depends on the ability to transfer skill on problem solving to the new problem. In this sense knowledge is subservient to skill or the servant of skill. At each level of the hierarchy of learning the student acquires new capacities for transfer. It is useful to think of this as a vertical transfer happening within the boundaries of the subject. But subjects (e.g., engineering) are divided up into many sub-disciplines (e.g., Acoustics, Aerodynamics, Thermodynamics) and students may find cognate transfer between these sub-disciplines difficult which makes it difficult for them to grasp real-life engineering problems [5]. Within subject or vertical transfer cannot be taken for granted. It cannot be left to chance or osmosis. For example, a study conducted by A. S. Luchins showed that students solving mathematical problems used the same problem solving strategy irrespective of whether it was the most appropriate strategy, an effect that he called set-mechanization [6]. In an action research in the classroom conducted by graduates training to be teachers it was found that transfer improved once the graduates had been shown examples of what was wanted [7].

Assessment has a key role to play in the development of within-subject transfer which it can help by the way the questions are designed. Given that transfer can only occur to the extent that students expect it to, wide ranging questions should help the student understand the new situation. Transfer will not be possible if the person seeking to understand a new situation does not understand the concepts and principles involved. It is incumbent, therefore, on teachers to ensure their students understand the concepts and principles to be learnt before they move their instruction forward [8]. Of equal importance is the fact that knowledge of models of problem solving and critical thinking will better enable students to handle new situations and develop reflective practice or meta-cognition. To put it in another way, students should benefit when they learn how to learn.

The transfer that is of more importance for the argument submitted here is what, for convenience, I used to call non-cognate or horizontal transfer. This occurs when a student is asked to move out from within-subject vertical learning to another subject that may or may not be cognate. In the case of mathematics and engineering that should be relatively easy but there are many examples of where in answer to problems set the mathematics is judged correct but the physical explanation is found to be incorrect [9]. As the cognate distance between subjects increases it seems that greater difficulty may be experienced with transfer. Brad Kallenberg calls this type of transfer cross-domain-transfer (CDT) [10]. It is useful to be aware of the cognate distance between two subjects, i.e., subjects that have close cognate connections compared with subjects whose cognate connections are apparently distant, although the cognate distance between subjects may not always be clear.

Kallenberg has argued that practical reasoning can be assisted if the reasoner looks outside the box whether it is solving an issue in engineering ethics or engineering design. In particular, he claims "that religious narratives can serve as one kind of source for CDT that lies outside of engineering proper. Complex problems are likely, more often than not, to contain cognate elements that are distant from the problem solvers cognate spectrum. The question is whether you see it?" Hence, the need to learn to see things from a distance. Elsewhere Dias has demonstrated a surprisingly close cognate relationship between history and engineering [11].

The transfer that is of particular interest here is that which has to be made when a person changes jobs that are apparently not closely related, and in particular to transfer from one mind set (way of thinking) to another [12]. Following the argument set out previously this may become increasingly necessary as the structure of work changes, a process that might be accelerated by the 2020 Covid-19 Pandemic. It may be that a person possesses generic skills that relate to a particular labour arena that have become shrouded by the experience of a particular job (see Section 6.2 [13]). To release them a person may need some training which, if it is to be effective, will be helped if, in his/her higher education, he/she has been trained in learning how to learn. It is not sufficient to leave students on their own to study and hope they become independent learners. Clearly, they will be better prepared if they have had a broad education because they may then have some familiarity, insight, into the job situation with which they are faced.

Some individuals may find cross-domain transfer easier than others. For example, Trevelyan argues that every student should do some teaching of their peers during their courses because the activity of teaching employs skills that are also necessary for "technical liaison" in industry [14].

Trevelyan says "first the engineer describes what needs to be done and when (the process of lesson planning), and negotiates a mutually agreeable arrangement with other people (students), who will be contributing their skills and expertise. Next, while the work is being performed, the engineer (teacher) keeps in contact with the people (students) doing the work to review the work and spot misunderstandings or differences of interpretation. The engineer (teacher) will also join in discussions of unexpected issues that arise and may need to compromise on the original requirements (task objectives). Third, when the work has been completed, the engineer (teacher) will carefully review the results (assess) and check that no further work or rectification is necessary." One might add that an effective engineer (teacher) will continually evaluate his/her own performance (self-accountability), a skill which will become increasingly important as more individuals work from home. Some may find this difficult but even those who find it easy may not appreciate that it is an exercise in the transfer of skill unless they are told. (The brackets are mine: for detailed descriptions of lesson planning, see [15]).

In response to Tevelyan I have suggested that there is much to be gained if students are able to teach in high school or college a subject like technological and engineering literacy [16]. Moreover, I also argued following induction, training should run concurrently with the teaching which not be an occasional exercise, particularly if skills in planning, classroom, management, and participant evaluation are to be developed and become transferable.

Nevertheless, the ability to effect transfer requires skill in learning. It is contended here that it is not sufficient to provide an environment that is conducive to learning but that individuals should through refection come to realise how they learn, how they might improve that learning, and how they might capitalise on it so that after higher education they are able to recognise what they need to learn, and plan and implement it. Furthermore, as was indicated in Chapter 5, they need to be able bring their skill in learning to understand the organisation and the way people interact in that organisation in which they will work. As Russell Korte indicated (Chapter 5) students who go on an industrial experience need to learn the political game that I played in organisations. One of the ways of achieving this would be to provide for an initial course in learning that embraced what is commonly called organisational theory. For example, understanding the power of the informal curriculum on learning should readily transfer to a search for the informal organisation.

The ability to transfer is also strongly influenced by the breadth of the curriculum and the extent to which it is arranged to provide for that enlargement of the mind that John Henry Newman argued was the goal of higher education (see Section 6.3).

6.2   THE RESPONSIBILITY OF EMPLOYERS

If schemes like the one suggested are to work employers have to believe that employees are likely to have more skills available to them than is apparent from their job description. This is the value of records of achievement or career portfolios [17]. In the past employers, particularly in Britain, have been reluctant to take redundant persons from engineering jobs when they do not perceive the person to have the specific skills they need. Coupled with the need to acquire a new work identity, this can lead to a self-fulfilling hypothesis where employees' come to believe they are only suitable for work in the areas for which they have been specifically trained as indicated by their job titles.

Is this "*occupational transfer gap*" real or imagined? I suggest that it is imagined, and that it is a perceptual problem. Most employees are likely to have skills that cross the divide of job perceptions, for which reason we need to have a skills approach to defining jobs [18] that is accompanied by educational programmes that encourage the possession of wide ranging interests and knowledge.

Thomas and Madigan identified this problem when they studied what happened to workers made redundant from the aircraft industry [19]. They suggested that there is need for employees and employers to think in terms of labour arenas. Youngman, Oxtoby, Monk, and Heywood when developing a new approach to task analysing engineers at work conceived of a labour arena as being a group of skills already possessed or which may be readily acquired [20]. It should enable the divide created by perceptions derived from job titles to be crossed to other less cognate jobs. They suggested that their technique of job analysis could be used to identify such arenas. In the model of higher "permanent" education proposed, the task of assessment is to identify the skills a person possesses, to identify weaknesses and remedy them, as well as to identify other skills that might be developed.

The model requires that employers who expect their employees to have to change jobs, even to having to leave the firm, take responsibility for the development of personnel, and in particular motivating them to learn. They will help themselves and their employees if they regard their organisation as a learning system. The concept of organisations as learning systems was popularised by P. M. Senge [21]. His model differs from that described by Youngman and his colleagues which shows the organisation in problem solving and decision making modes (Exhibit 6.1). The learning curves are similar to Richard Foster's "S" curves which describe the effort put into improving the product or process, and the results the company obtained from investment (see Exhibit 6.1 (i)) [22]. It will be seen that this curve parallels the learning curves in Exhibit 6.1 (ii) and that each new adaptation (Exhibit 6.1 (i-b; ii-d)) represents an additional cost. In Foster's model a new-product process starts a new S curve (Exhibit 6.1 (b)), and in this respect Foster's model adds an important dimension to the model. The curves in Exhibit 6.1 (ii) show that from the acquisition of an idea by the company goes through the same processes of learning as individuals, except that the phases take place over a long period of time. As I argued above, the factors that enhance or impede learning also impede or enhance innovation in the organisation [23]. Such organisations are communities of learning (see Section 6.5). That learning is enhanced if it is "enlarged."

6.3   ENLARGEMENT OF THE MIND AND ADAPTABILITY

The requirement that we should become more adaptable makes its own case for a broad education. However, it does not dictate what that breadth should be or how it should be structured. It does mean that however the curriculum is structured, it should focus on what John Henry Newman called "enlargement or expansion of mind" [24]. For Newman enlargement like knowledge was a process through which a person obtained a philosophical disposition or wisdom. Therefore, one of the aims of higher education is to help students acquire the knowledge and skills necessary to enlarge their minds. Often, when he is writing about what today are called the outcomes of university education, Newman uses terms and phrases from which we can only conclude that he embraced all those domains of the person that are non-cognitional, sometimes called "affective." For example "such an intellect [...] cannot be impetuous, cannot be at a loss, cannot but be patient, collected, and majestically calm" [25]. It is to be found in his idea of a university tutor (teacher), and the value he placed on residence (in hall). For us, it is a reminder that the skills of the affective domain are as important as those in the cognitive domain. This is the point that industrialists are making when they demand that graduates possess personal transferable skills (see Chapter 5).

Newman argued that "knowledge itself, though a condition of the mind's enlargement, yet whatever be its range, is not the very thing that enlarges it" [26]. Rather, it is the ability

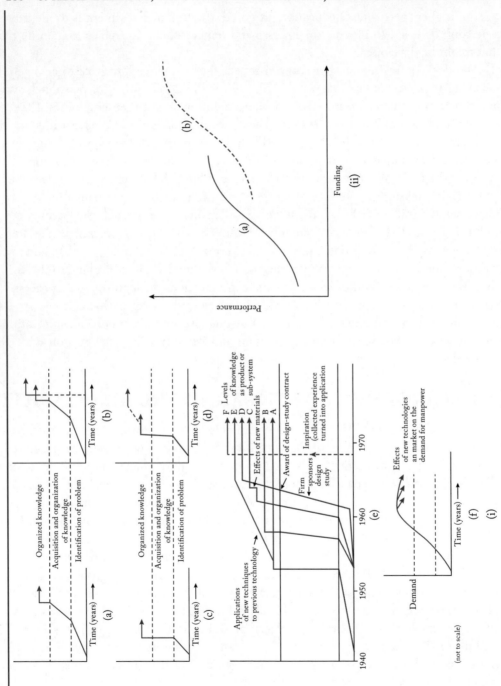

Exhibit 6.1: (i) Shows the pattern of innovation in the firm showing process from problems identification to problem solved, feedback and the next curve in the process of development and so on; (ii) illustrates the rate of change in demand for specific types of workforce as a function of the learning/innovation curve.

to perceive the relationships between subjects. "Enlargement consists in the comparisons of the subjects of knowledge one with another. We feel ourselves to be ranging freely when we not only learn something but when we also refer it to what we know before" [27]. This would seem to be consistent with that present-day view of learning that considers it to be the process by which experience develops new and reorganises old responses [28]. This is clearly what happens or should happen within courses. Without it there could be no development or movement within a course. But demonstrating that knowledge has been acquired is no guarantee that there has been enlargement.

Adaptability arises from a person's ability to think and reason. It is these abilities applied across a range of subjects that enlarge the mind. The one thing in which we are all engaged is reasoning. We are all engaged in "deducing well or ill, conclusions from premises, each concerning the subject of his own particular business"—"The man who has learnt to think, and to reason and to compare, and to discriminate and analyse …will not at once be a lawyer […], or a physician, or a good landlord, or a man of business, or soldier, or engineer, or chemist […] but he will be placed in that state of intellect in which he can take up any one of the sciences, or callings I have referred to, or any other for which he has a taste or special talent […]" [29]. That is the essence of an educated person.

But in today's understanding "transfer" will not take place if these subjects are taught independently of each other [30]. Since transfer will only occur to the extent we expect it to occur, the curriculum has to show how it can occur in what might best be described as interdisciplinary or trans-disciplinary situations. The failure to approach study in this way is the reason why a general education that comprises the study of a number of independently organised subjects is not liberal. It is the reason why in subject specialisms like engineering so many students are unable to combine knowledge from the sub-disciplines to solve complex problems. Taught in a way that overcomes this problem, that is, in a spirit of universality, engineering is as much a liberal study as any other [31]. Any detailed analysis of the activity (process) of engineering will demonstrate that this is so.

6.4   EXPERTISE AND LIBERAL EDUCATION

In 2016, the results of the referendum in the UK and the election for President in the United States shocked their respective establishments [32]. Immediately it was argued in the UK that the populists were illiterate. Graduates were found to have voted in large numbers for the losing side. Moreover, the losers found it difficult to understand why the views of large numbers of so called experts had been rejected. The assumption that because a person has a degree he/she has some general intellectual advantage over a person who has not may be questioned. The most that can be said of that person is that he/she has an academic advantage. They may well have a higher intelligence quotient but that is not necessarily accompanied by practical intelligence or wisdom. There are many examples of expert views being found wanting, and expertise was questioned by some politicians who in other circumstances would regard themselves as experts! [33]. Be that as it may, the greater problem is the notion put forward by politicians, apparently unconscious of the contradiction that political decision making should be left to experts. The Covid-19 Pandemic showed that experts can disagree, particularly when data is not safe or missing.

That said, there will be general agreement to the assertion that every person in a democracy should be able to engage in political debate irrespective of educational attainment. It is also likely to be agreed that the level of debate will be raised if they have had a high level of liberal education as defined by Newman. That is the second, if not more important point in support of a liberal education, the object of which is to give the student a unified vision of reality. The point of a University is to make available the widest possible range of studies which while students will not be able to study all of them, "they will be the gainers by living among those and under those who represent the whole circle" (of knowledge) [34, p. 101].

"This I conceive to be the advantage of a seat of universal learning, considered as a place of education. An assemblage of learnt men, zealous for their own sciences, and rivals of each other, are brought, by familiar intercourse, and for the sake of intellectual peace, to adjust together the claims and relations of their respective subjects of investigation. They learn to respect, to consult to aid each other. Thus, is created a pure and clear atmosphere of thought, which the student also breathes, though in his own case he only pursues a few sciences out of the multitude" [35, p. 101].

It is the principle of the collegiate university. Thus, "when a multitude of young men, keen and open hearted, sympathetic and observant, as young men are, come together and freely mix with each other, they are sure to learn from one another, even if there be no one to teach them […] A parallel teaching is necessary for our social being, and it is secured by a large school or a college; and this effect may be fairly called in its own department an enlargement of mind" [36, p. 146].

Fast forward 100 or so years and the UK Ministry of Education is found to have suggested how technical colleges might introduce liberal studies in their colleges [37]. Those CATs that wished to offer Diploma in Technology courses were required to include programmes of liberal study (see Chapter 5). The reason for not requiring technology students in universities to undertake programmes of liberal study, apart from the fact that universities were regarded as private self-governing institutions, seems to have been that the student environment they provided was liberal, and the views cited above supported their case. But that was to misunderstand or misuse Newman.

In 1977, Alexander Astin published a major study of the impact of liberal arts colleges on students in the United States. It was titled, "*Four Critical Years*." When he published a revised version in 1993 he added to the title "*What Matters in College*" [38]. While Newman is not referenced the results give considerable support to his views. Indeed, Astin specifically supports the Oxbridge model. He writes, "This study has shown, once again, that this traditional model of undergraduate education leads to favourable educational results across a broad spectrum of cognitive and affective outcomes and in most areas of student satisfaction. Perhaps most important, however, is the finding that institutional structure, as such, is not a key ingredient; rather it is the kinds of peer groups and faculty environments that tend to emerge under these different structures" [39, p. 413]. The similarities with the findings of the Hamilton College study will be apparent (Section 3.6, [40]). The problem is how to arrange the learning environment to achieve these goals. Related to this discussion is the question, "Can the circumstances of small group learning be repeated in e/distance learning?"

6.5   COMMUNITY AND LEARNING

John Macmurray the Scottish philosopher distinguished the notion of "society" (a cooperative enterprise maintained by justice and a harmony of functions) from the concept of "community" (the full expression of their togetherness by members of a society—in personal communion through culture) [41]. Summarizing Macmurray's views on education given in the Payne lectures to the College of Preceptors Costello writes, "Education [...] must aim to serve both realities at once but with a vision that situates the functional, social goal (learning skills and aptitudes) as a subordinate dimension within the cultural one (personal formation and development in community). These are not two separate kinds of education but two aspects of the same education process [...] It is impossible to teach any technical growth whatever without producing some cultural effect. Equally it is impossible to enhance expression without stimulating growth in technical competence. But the latter should be integrated within the former and directed to its service. In other words, every growth in technical know-how should be taught in the context of responsibility—to people and to our culture" [42], and I would add to ourselves. Some might consider that universities have a major obligation to ensure they function as a community given that technology has substantially reduced shared experience and debate which, it has been argued, is a major cost to society [43].

In terms of the thesis offered here, it provides for the reconciliation of the personal and professional identities.

Macmurray would have agreed with Albert North Whitehead that "there is only one subject matter for education, and that is life in all its manifestations" [44]. If we are concerned with learning about life in all its manifestations then we are likely to be more adaptable and flexible and to quote John Henry Newman "to fill any post with credit, and to master any subject with facility" [45].

There can be no contradiction between a liberal education properly constructed so that a person experiences an enlargement of mind and the demands of industry. Those demands when analysed are of two kinds. The first is for specific kinds of knowledge, some of which could quite evidently should be provided by industry. The second is for interpersonal skill which necessarily embraces the "affective," a point which is illustrated by the "person centred approach adopted by the Viennese researchers Renate Motschnig-Pitrik and Katherine Figl for the development of these skills among computer scientists" [46].

As I have argued in other papers Newman's statement of the outcomes of a liberal university education is entirely consistent with the aims of education proposed by MIT in *Made in America. Regaining the Productive Edge* [47], or the UK Employment Department's statement of learning outcomes in *Enterprise Learning and its Assessment* [48], or more significantly Sternberg's list of abilities that contribute to intelligence in *A Triarchic View of Intelligence* suggested by experts and lay persons [49]. But, in contrast, Newman's "idea" begins with view that a "person" is something very much more than a cognitive processing machine. Newman's philosophy is enriched by Macmurray's understanding of how we become a person.

Persons only develop as persons in relation to other persons. We come to be who we are as personal individuals only in personal relationships. It is that view which is the justification for the collegiate organisation of a university. It is that view which gives credence to cooperative learning. Nevertheless, reason seems to be withheld, even in the face of evidence. One reason for this is the socio-cultural history of the curriculum of a particular country or culture [50].

6.6   CONCLUSION

If it is correct that many individuals will have to change jobs on a number of occasions during their lifetime then industrial organisations will have to accept that they share with society, (probably represented by an institution of higher education) responsibility not only for the professional but the personal development of the individual. Again, if individuals are likely to work for much less time than they have done in the past as some forecasters predict the personal begins to become more important than the professional. That would be to correct the present imbalance between the professional and the personal in the curriculum; for, it is the "personal life" that is the driver of all our behaviour.

A curriculum has to reflect the personal as much as it does the professional and society. The demands of life and work will require individuals to become more adaptable than in the past and to be able to transfer to situations that are outside their zone of comfort. This will require highly developed skill in learning which is currently not provided by higher education institutions. Organisations will benefit because the most effective organisations are those that learn. The ability to cross-domain transfer is enhanced if a person has a broad-based education on which they can build the specialist studies they will need from time to time. Apart from Kallenberg, a theologian [51], other engineering educators have argued that philosophy should be a component of the engineering curriculum partly because as Grimson has argued engineering is fundamentally philosophical in nature [52], or as Smith and Korte have argued the skills of philosophical inquiry are of immense value to students [53], or that it is an essential component of the liberal education of engineers [54]. As Grimson and his colleagues argue engineers should be able to act in the role of public intellectuals not just as technocrats [55]. To achieve that end they need to acquire a philosophical disposition and that requires learning something more than just engineering.

6.7   NOTES AND REFERENCES

[1] Trist, E. and Bamforth, W. (1951). Some social and psychological consequences of the long wall method of coal cutting. *Human Relations*, 4:3–38. 95

See also Emery, F. (1969). In (ed.), *Systems Thinking*. Harmondsworth, Penguin.

[2] Heywood, J. (1989). *Learning, Adaptability, and Change. The Challenge for Education and Industry*. London, Paul Chapman/Sage. 95, 105, 106

This argument should not be taken to underestimate the positive value of tacit knowledge in everyday life. Indeed it is often underestimated.

[3] Sternberg, R. S. (1985). *Beyond IQ. A Triarchic View of Intelligence*. New York, Cambridge University Press. 95, 110

[4] *Loc. cit* Ref. [2]. 95

[5] Culver, R. S. and Hackos, J. T. (1982). Perry's model of intellectual development. *Engineering Education*, 73(2):221–226. 96

Culver, R. S. and Olds, B. (1986). "EPICS" an integrated program for the first two years. *Proc. of the St. Lawrence Section Conference, 1986*. American Society for Engineering Education.

[6] Luchins, A. S. (1942). Mechanisation in problem solving: The effect of "Einstellung." *Psychological Monographs*, no. 248 (it is sometimes called set induction). 96

[7] Heywood, J. (1992). Student teachers as researchers of instruction in the classroom in J. H. C. Vonk and H. J. van Helden (Eds.). *New Prospects for Teacher Education in Europe*. Free University of Amsterdam and the Association of Teacher Education in Europe, Brussels. 96

[8] Saupé, J. (1961). Learning in P. Dressel (Ed.). *Evaluation in Higher Education*. Boston, MA, Houghton Mifflin. 96

[9] Glyn Price cited by Heywood, J. (1989). *Assessment in Higher Education*, 2nd ed., Wiley, Chichester. 96

[10] Kallenberg, B. J. (2013). *By Design. Ethics, Theology, and the Practice of Engineering*. Cambridge, UK, James Clarke. 96, 109, 110

[11] Dias, P. (2013). The disciplines of engineering and history: Some common ground. *Science and Engineering Ethics*. Springer. 97, 109

[12] Heywood, J. (2007). "Think… about how others think." Liberal education and engineering. *ASEE/IEEE Proc. Frontiers in Education Conference*, T3C:2–24. See Chapter 2 Heywood, J. (2017). *The Human Side of Engineering*. San Rafel, CA, Morgan & Claypool. 97

[13] Youngman, M., Oxtoby, R., Monk, J. D., and Heywood, J. (1977). *Analysing Jobs*. Aldershot, Gower Press. 97, 106

[14] Trevelyan, J. (2010). Engineering students need to learn to teach. *ASEEE/IEEE Proc. Frontiers in Education Conference*, F3H:1–6. 97

[15] Heywood, J. (2008). *Instructional and Curriculum Leadership. Towards Inquiry Oriented Schools*. Dublin, Original Writing for National Association of Principals and Deputies. 97

[16] Heywood, J. (2015). Teaching, education, Engineering, and technological literacy. *Proc. Annual Conference American Society for Engineering Education*, paper 1290. 97

[17] Heywood, J. (2000). *Assessment in Higher Education. Student Learning, Teaching, Programmes, and Institutions*. London, Jessica Kingsley. 98

[18] *Loc. cit*. Ref. [13, p. 93]. 98

[19] Thomas, B. and Madigan, C. (1974). Strategy and job choice after redundancy. A case study. *Sociological Review*, 22:83–102. 98

[20] *Loc. cit*. Ref. [13, p. 93]. 98

[21] Senge, P. (1990). *The Fifth Discipline. The Art and Practice of the Learning Organization*. New York, Doubleday. 99

[22] Foster, R. (1986). *Innovation. The Attacker's Advantage*. London, Macmillan. 99

[23] *Loc. cit*. Ref. [2]. 99, 109, 110

[24] Newman, J. H. (1852). *The Idea of a University* (with additional lectures added in 1873. 1923 Impression), pages124–150, London, Longmans, Green. Discourse VI. Newman did not conceive of enlargement being caused by a wide range of knowledge though the study of a large number of subjects […] "Liberal education is not mere knowledge, or knowledge considered in its *matter* […] whether knowledge, that is, acquirement, is after all the real principle of the enlargement, or whether that principle is not something beyond it," p. 130. For a commentary see Culler, A. D. (1955). *The Imperial Intellect. A Study of Newman's Educational Ideal*, pages 204–208, Yale University Press. The principles of Newman's theory of knowledge will be found in his University Sermons especially no. 14. 99, 109

[25] *The Idea*, p. 138. Perhaps the most quoted paragraphs on this point end the seventh discourse pages 177 and 178 of *The Idea*. They read "University training is a great ordinary means to a great ordinary end: it aims at raising the intellectual tone of society, at cultivating the public mind at purifying national taste, at supplying true principles to popular enthusiasm and fixed aims to popular aspiration at giving enlargement and sobriety to the ideas of the age, at facilitating the exercise of popular power, and refining the intercourse of private life. It is the education which gives [persons] a clear conscious view of [their] own opinions and judgements, a truth in developing them, an eloquence in expressing them; and a force in urging them. It teaches [them] to see things as they are, to go right to the point, to disentangle a skein of thought, to detect what is sophistical; and to discard what is irrelevant. It prepares [them] to fill any post with credit, and to master any subject with facility. It shows [them] how to accommodate himself to others, how to throw himself into their frame of mind, how to bring before them his own, how to influence them, how to come to an understanding with them, how to bear with them. [They] are at home in any society, [they] know when to speak and when to be silent, [they are] able to converse, [they are] able to listen, [they] can ask a question pertinently, and gain a lesson seasonably, when [they] have nothing important [themselves], [they are] ever ready, yet never in the way, [they are] a pleasant companion, and a comrade you can depend upon; [they know] when to be serious and when to trifle, and they have sure tract which enables [them] to trifle with gracefulness and to be serious with effect. [They] have the repose of mind which lives in itself, and which has resources for its happiness at home when it cannot go abroad. [They have] a gift which serves them in public and supports [them] in retirement, without which failure and disappointment have a charm. The art which tends to make a [person] all this, is the object which it pursues as useful as the art of wealth or the art of health, though it is less susceptible of method, and less tangible, less certain and complete in the result." 99

[26] Newman, J. H. (1890). *Fifteen Sermons Preached Before the University of Oxford*, 3rd ed., London, Rivingtons, 14th Sermon para. 21, p. 287. 99

[27] *Ibid.* 101

[28] Saupé, J. L. (1961) in Dressel, P. (Ed.). *Evaluation in Higher Education*, Houghton Mifflin, Boston, MA, or see Heywood, J. (1982). *Pitfalls and Planning in Student Teaching*, p. 14, London, Kogan Page. 101

[29] *The Idea*, p. 166 in Discourse VII Knowledge and Professional Skill. 101

[30] Transfer is related to critical/reflective thinking. See Heywood, J. (2000). *Assessment in Higher Education. Student Learning, Teaching, Programmes, and Institutions*, pages 177–181 and 187–191, London, Jessica Kingsley. Section on the critical thinking movement in higher education. 101

[31] The argument was set out differently, and in a little more detail in my paper (1505) on *Engineering Literacy: A Component of Liberal Education*, to the technological Literacy Division at the *Annual ASEE Conference 2010*. Under the heading, "implications for the curriculum" it reads—Newman considered that all "[...] the branches knowledge are, at least implicitly the subject matter" of the university curriculum. Imparting this knowledge in a philosophical way gives a university its culture. This "culture is a good in itself; that the knowledge which is both its instrument and result is called Liberal Knowledge; [...]" (a) It is a universal knowledge and the university is conceived of as a *Studium Generale* or a school of universal learning. 101

The issue is the meaning of "universal." Today we take it to mean "all inclusive" but as Culler pointed out this was not the case in Newman's time especially as it was used in the context of university education. At that time its usage derived from *"uni-versum"* and meant *"turned into one."* There was, Culler writes—"a desire to see things whole that forced men to look at the whole body of things, and therefore the true character of a university is not that it teaches all the sciences but that whatever sciences it does teach, it teaches in a spirit of universality." (b) But this does not mean that it can be done within a single specialism for each subject has something of its own that is specific to itself to offer. Newman wrote; "If we might venture to imitate [...] Lord Bacon, in some of his concise illustrations of the comparative utility of the different studies, we should say that history would give fullness, moral philosophy strength, and poetry elevation to the understanding [...] the elements of good reason are not to be found fully and truly expressed in any one kind of study [...]" moreover, "if different studies are useful for aiding, they are still more useful for correcting each other [...]." (c) Each study has its own characteristic way of thinking, and each subject makes its own contribution to our understanding, hence the importance of the skill of thinking. It is the argument for including engineering as a subject in the curriculum of general education.

Necessarily "man," who is at the centre of this aim has to be viewed in all his relationships [...] "What is true of man in general would also be true of any portion of reality however minute. If we wished to know a single material object—for example, Westminster Abbey—to know it thoroughly, we should have to make it the focus of universal science. For the science of architecture would speak only of its artistic form, engineering of its stresses and strains, geology of its stones, chemistry and physics of the ultimate constitution of its matter, history of its past and literature of the meaning which it had for the culture of a people. What each one of these sciences would say would be perfectly true to its own idea, but it would not give us a true picture of Westminster Abbey." (d) So wrote Culler to further illustrate Newman's idea. To get a true picture the sciences would have to be recombined and this recombination is the object of university education. We might go further and add that it is through such recombinations that advances in thought and practicalities are made. But Newman did not think recombination was the same as all the subjects taken together. It "is a science distinct from them all and yet in some sense embodying the materials of them all." (e) This activity is what Newman called liberal knowledge and at other times, as Culler notes, philosophy, *philosophia prima*, architectonic science or science of the sciences. He did not pursue this concept in any great detail, but in the today's jargon it would seem to be a reflective activity of synthesis; an ability to bring all the parts together in order to make a judgement, for which reason the subjects of the curriculum cannot be taught as entities isolated from each other. The consequences of this capability for the educated person so produced were set down by Newman in the oft quoted statement about the ends of a university education given in Ref. [10] above. This statement clearly shows the importance of development in the "affective" as well as the "cognitive domain." To gain such a comprehensive view, study of as wide a range of knowledge as is possible, this to include engineering, is necessary; but how is that to be achieved without some missionary activity and curriculum innovation, and at what level?

(a) *Loc. cit.* Ref. 134, p. 214. Discourse IX. Duties of the church towards knowledge.

(b) Culler, A. D. (1955). *The Imperial Intellect. A Study of Cardinal Newman's Educational Ideal*, p. 180, New Haven, CT, Yale University Press.

(c) *Loc. cit.* Ref. 139. p. 175.

(d) Culler's illustration is chosen over Newman's because of the reference to engineering, p. 182, Ref. [11].

(e) *Loc. cit.* Ref. [24] (Heywood, 2005) p. 182. In Newman Ref. 139 see p. 43 ff of the second discourse and the fifth discourse.

[32] (a) Zitner, A. and Chinni, D. (2016). Voters education level a driving force. *The Wall Street Journal*, October 14. 101

(b) Ferguson, N. (2016). This was no whitelash, it was a vote to get America working. *The Sunday Times*, November 13.

[33] Gove, M. (2016). "Experts" like Carney must curb their arrogance. *The Times*, October 21. 101

[34] *Loc. cit.* Ref. [23]. 102

[35] *Ibid.* 102

[36] *Ibid.* 102

[37] (i) Davies, L. (1965). *Liberal Studies and Higher Technology*. Cardiff, University of Wales Press. 102

(ii) Heywood, J. (2011). Higher technological education in England and Wales, 1955–1966. Compulsory Liberal studies. *Proc. Annual Conference of the American Association for Engineering Education*, Paper 635.

[38] Astin, A. W. (1993). *What Matters in College? Four Critical Years Revisited*. San Francisco, Jossey-Bass. 102

[39] *Ibid.* 103

[40] Chambliss, D. F. and Takacs, C. G. (2014). *How College Works*. Cambridge MA, Harvard University Press. 103

[41] Macmurray, J. (1961). *Persons in Relations*. London, Faber and Faber. 103

[42] Costello, J. E. (2002). *John Macmurray. A Biography*, page 316 ff, Edinburgh, Floris books. 103

[43] Lucey, C. (2017). Downside of diversity is entrenched division. *The Times*, September 1. 103

[44] Whitehead, A. N. (1932). *The Aims of Education*, 2nd ed., (Chapter 1), London, Ernest Benn. 103

[45] *Loc. cit.* Ref. [23, p. 131]. 103

[46] Motschnig-Pitrik, R. and Figl, K. (2007). Developing team competence as part of a person centered learning course on communication and soft skills in project management. *Proc. Frontiers in Education Conference*, F2G:15–21. 103

[47] Dertouzos, M. L., Lester, R. K., Solow, R. M., and the MIT Commission on Industrial Productivity (1989). *Made in America. Regaining the Productive Edge* (Chapter 12), Cambridge, MA, MIT Press. 104

[48] Heywood, J. (1994). Enterprise learning and its assessment. *Technical Report No. 20*, Exhibit 1.4, p. 17, Sheffield Learning Methods Branch, Employment Department. 104

[49] *Loc. cit.* Ref. [3]. 104

[50] See for example, Ellis, W. (1995). *The Oxbridge Conspiracy. How the Ancient Universities have Kept Their Stranglehold on the Establishment*. London, Penguin, and Stevens, R. (2004). *University to Uni. The Politics of Higher Education in England since 1944*. London. Politico (Methuen). 104

[51] *Loc. cit*, Ref. [10]. 104

[52] (a) Grimson, W. (2007). Engineering—An inherently philosophical enterprise. In Christensen, S. H., Meganck, M., and Delahousie, B. (Eds.). *Philosophy in Engineering*. Aarhus, DK, Academica. 104

(b) Grimson, W. (2104). Engineering and philosophy in Heywood, J. and Cheville, A. (Eds.). *Philosophical Perspectives on Engineering and Technological Literacy*. A Handbook of the Technological Literacy Division of the American Society for Engineering Education Washington DC, American Society for Engineering Education.

[53] Korte, R. and Smith, K. (2009). Developing engineering student's philosophical inquiry skills. *ASEE/IEEE Proc. Frontiers in Education Conference*, T4b:1–2. 104

[54] Grimson, W., Dyrenfurth, M., and Murphy, M. (2009). Liberal studies in engineering and technology in Christensen, S. H. Delahousie, B., and Meganck, M. (Eds.). *Engineering in Context*. Aarhus, Dk, Academica. 104

[55] *Ibid.* 104

CHAPTER 7

Society and Technology

7.1   THE AIMS OF HIGHER EDUCATION

The purpose of Chapters 3–6 has been to illustrate the process of curriculum design (development) based on the idea of screening put forward by Ralph Tyler and E. J. Furst. This was achieved by re-examining the curriculum and structure of higher education in light of (a) the increasing cost of higher education and (b) rapidly changing technology on the structure of employment. These studies embraced both the UK and the U.S. and took into account the different cultural conditions in which their sub-systems of higher education function.

One way of reducing costs is to reduce the length of the course (Section 3.6). Traditionally, undergraduate, full-time degree courses in the UK have been of three year duration and in the U.S. of four year duration with students in classes for just over half the year. Normally students entering university in the UK are expected to complete their degree within the three-year period allocated for study. To become a chartered engineer, a four-year course is required. No good reasons for these structures have been provided and it would be easy to re-organise the academic year so as to compress three years into two, and in the U.S. four years into three. In the UK it was estimated that this would reduce the costs of tuition by 20% in consequence of which universities are being encouraged to run two-year programmes (Section 3.6). In both countries it is difficult to understand why it is that three- and four-year programmes have become the norm except in so far as the curriculum generally, and within subjects specifically. It is seen as necessary for the quantity of information considered to be required to be given. Irrespective of the several ideologies of instruction at play within the engineering curriculum, the prevailing epistemology is that of knowledge as information. There is little or no attention paid to the ontological dimension of higher education, that is, its concern with the individual as a "being" and therefore, with the personal.

Recommendations to compress courses retain this philosophical outlook. This is what some universities are doing. Yet the study of the impact of technology on the structure of work, and subsidiary studies of employer's attitudes to higher education suggest that there is a need to radically rethink higher education. For example, the criticism that graduates are not equipped with the "soft" or "professional" skills cannot be met by simply adding apparently non-cognate subjects to the curriculum. In any case, this "inside" view of competency that is, the belief that the competencies required reside within the individual and can, therefore, be taught has been found wanting (Section 5.4). Work place, indeed life competencies, depend in no small way on the context in which they are developed. This applies to adaptability as much as to anything else.

An employer can demand employees be adaptable but, at the same time, arrange the organisation so as to impede adaptation, unconsciously if not deliberately.

The trend among employers to accept as little responsibility for their employees as they can, coupled with an increasing rate of redundancy of knowledge, means that employees have to put themselves in a position that when faced with the possibility of unemployment they are prepared to move into activities that are non-cognate with their current knowledge and skill. Changes in the structure of work brought about by technology will mean that the average person will have to make several career changes during their working life. Some of these changes may require cross domain transfer. This will require individuals not only the motivation to transfer across domains but skill in devising programmes of learning which will enable them to move forward with their careers. Therefore, a basic aim of higher education should be to equip graduates with skill in independent learning, and the associated quality of reflective practice. Such skill will enable the graduate to gain new specialisms and expertise.

Clearly, if one is to practice cross-domain transfer one has to experience it. Such experience may be provided by a liberal (as opposed to a general) education (Chapter 6). It does not require students to study every subject under the sun but it does require them to be prepared to undertake in-depth study in an area that is currently non-cognate, and be able to pursue that subject on its own (independent learning). The discovery by the student of the independent learning that a student wishes to undertake may be by means of a problem-based course that enables the student to explore a variety of learning contexts relevant to both his personal life and work.

The fundamental aim of the curriculum now emerges as a "preparation for and continuing support for life and work." As such, it would have to acknowledge that living takes place within a technological society. That would demand in every day parlance that its products be techno-logically literate. In contrast to specialist and generalist subject models of higher education, an alternative two-year model is proposed that provides for a basic input on learning which is built on throughout the course. Such a course would be accompanied by periods of industrial prac-tice in order for students to experience different ways of knowing, together with an academic a course structured around problem- and project-based learning that leads to the design and personal implementation of independent learning. Lifelong learning is achieved by the student retaining a link with the university that would enabled him to purchase packages of substantive study designed by him/herself from the university that may or may not require attendance at the institution.

The purpose of this chapter is to propose an outline of a model for a two-year programme to meet the requirements of a "basic" higher education as a preparation for further learning, living and working in an advanced technological society. That is, a person who is technologically literate. It is a goal that can be achieved in any number of ways. The curriculum derived here is derived from screening the material discussed in the previous chapters. This chapter is intended to derive a framework (or model) not a detailed description of content and method.

7.2   PROSPECTS: DETERMINING THE PROGRAMME

All of us face a world that is considerably challenging. Without thinking it is possible to see eight challenges that stare us in the face viz climate change; longer living (care of the elderly); the reform of capitalism; war and pestilence; global inequalities; information overload; impact of the internet and impact of technology; individuals will have their own views of their relative importance in the scheme of things.

They have in common that the solutions to problems in each of these challenges require knowledge from numerous areas of knowledge (i.e., the disciplines). Since in each of these areas of challenge technology can both contribute to and solve problems, a curriculum that is focused on Society and Technology is fully justified. Given the impact of technology on our lives it will be necessary to provide appropriate studies in technical proficiency throughout the programme. Problems set in any of these areas of "challenges" will require information from a variety of subjects. Inspection of Exhibit 1.2 and its associated text (Section 1.5) shows that among the subjects that would have to be studied ("I know") would be engineering design, engineering science, economics, industrial psychology, law, manufacturing, sociology, and philosophy (especially ethics), and these without mentioning the practical ("I do") element that any programme should have. The list would change if technology is being used to solve a problem in one of the other challenge areas.

Technological change impacts us in very different ways. This can be seen from the influence of electro-mechanical devices in the kitchen and the impact they have had on families, or at a societal level, or for example, in the concern that communities have for processes like *fracking* which enable large quantities of gas to be extracted from shale. At the micro-level, changes in kitchen technology or for that matter electronics in the home are adopted to without much thought. Not so when a community is faced with *fracking*. Similarly, while cell (mobile) phones were quickly accepted, some communities reacted unfavourably to the erection antennas to provide the service because of fears of cancer. To persuade communities to adopt an unwelcome technology engineering explanations for the acceptance of the technology will be given by experts or company officials which the plaintiffs will be expected to take on trust. For the plaintiffs to make a rational decision they will have to have some knowledge of "risk" and some understanding of the engineering together with an ability to synthesise the two. In a word, they will need to be engineering and technologically literate if they are to have any control over the situation. The curriculum problem, some would say conundrum, is "what do we need to know to judge with reasonable competence whether or not we can trust the expert(s)?"

It is reasonable to argue that we have been very naïve in our acceptance of technology in the home. We now find that we do not know how to control its more insidious effects. How do we prevent children from becoming the victims of pornography or, even worse, grooming? How do we judge what is fake and what is not, truth from fiction and so on? It is evident that technology is controlling us more and more. Thus, while there is no need for a detailed understanding of technology to resolve this issue there is clearly a need to understand our value system as well as those of others. So any program in technological literacy has to function at different levels of apprehension within different dimensions of knowledge ranging from the technological and scientific to the philosophical and personal.

A traditional subject-based curriculum would face the problem of having at one end of the curriculum spectrum traditional engineering science-based courses that rely heavily on a knowledge of physics and mathematics, while at the other end would be programs in technology and society that rely on methods in the social sciences and philosophy, and make little or no reference to engineering processes. Such an arrangement would not fulfil the requirements for a liberal education except in so far as it insured that the interrelationships between the subjects (courses) that make up the programs were understood.

Alternatively, a program that is primarily structured around problem- and project-based learning should be able to find coherence through a framework of key concepts. At the same time such a framework would assist the development of skill in non-cognate (cross domain) transfer.

7.3   EXAMPLES OF "BRIDGING" KEY CONCEPTS

There are a number of concepts (bridging) that enable an individual to better understand the differences and similarities between different areas of knowledge. For example, the key concept "perception" (see Section 7.6). Key concepts in curriculum development constitute recurrent themes, the threads of which run throughout the curriculum in a cumulative overarching pattern [1]. For example as between the humane subjects, social sciences and science and technology the concepts, "cause/consequence," "conflict/consensus," "continuity/change," "evidence," "mistake," "discrepancy," "power," "probability," "random error/systematic error," "risk," "similarity/difference," "uncertainty," and "values and beliefs" have some level of meaning that is either strong or weak depending on the circumstance of its use. One way of developing a curriculum is to map the essential key concepts [2].

Paul Steif has shown how he has used "concept clusters" to organise both assessment and instruction in statics [3]. He distinguishes between (a) skills that are actions that can be mastered by rote practice and (b) concepts that demand much more careful explanation and deeper understanding. He argues that some student errors may stem from inadequate skills rather than conceptual misunderstanding. Related to this is an argument by Papadopoulos that procedural knowledge should also be emphasised if there is to be conceptual understanding [4].

Examining how key concepts are used (not used) in each subject (area) gives some idea of the differences between subjects. Thus, in a test in a physics for arts students course, the students were asked to "Compare the usefulness of the concept of error as used in physics with that of the errors occurring in the study of your major subject" [5, p. 85]. (Any of the concepts listed can be substituted for "error" in the question.) For example, students might be asked to consider the differences between engineering and science and the roles of "risk" and "failure" in them. Engineering texts tend to look at these two concepts within the context of how they improve engineering. Thus, Vincenti argues that for as many successes there are in engineering there are likely to be just as many failures [6, p. 46]. Davis considers that engineers and managers differ in the way they approach "risk." "Engineers" reduce risk to "permissible levels" whereas managers "balance risk against benefit" [7, p. 67]. In either case they do not seem to face up to the everyday problems posed by the consumer. Should we worry about an ophthalmic surgeon's capability when he gives a wrong scientific explanation for laser surgery, or should we go on his/her proven record of success with such surgery? [8, p. 212]. The obligations acquired with the right to vote mean that we have to understand the risks that governments take with our money.

This point is no better or powerfully illustrated than by the Covid-19 Pandemic. Governments have spent huge sums of money to enable businesses to be closed, and people to stay at home. At the same time real fear was spread among certain members of population that resisted "unlocking." They were risk averse but many others, particularly younger people, judged that there was not a great risk. One problem was the uncertainty some scientists, who, for example, disagreed about the wearing of facemasks, or whether or not it was necessary to vaccinate children.

7.4   THE "OVER-ARCHING" CONCEPT OF DESIGN

While these key concepts also act as bridges (bridging concepts) in so far as engineering is concerned they are but fragments of the process which are brought together by overarching concepts. David Perkins could claim that "design" is just such a concept [9] since his theory of knowledge is based on the principle that knowledge is design. This means that structures such as governments, theorems, experiments, and short stories are designs, which is to broaden the whole concept of what we perceive to be design. He challenges the assumptions that: (1) knowledge is information; (2) the view that you cannot think or do science (engineering) unless you know a lot of science (engineering); (3) understand something that is hard to understand for which reason "teachers can only give you lots of exercises relevant to such accomplishments;" and (4) standard problems and exercises capture the skills of a discipline. For example, it is assumed that "when you solve a textbook problem you are doing a large part of what being a chemist involves, albeit at an elementary level" [10, pages 212–213].

He makes a clear distinction between knowledge as information and knowledge as design. Knowledge is created as a result of design; it is also understood as design. "One might say that a design is a structure adapted to a purpose" [11]. That knowledge is adapted so that it can be transferred to other situations. An important ability in the designer's repertoire is to be able to break away from experience or, as Perkins puts it, "familiar frames of reference."

Knowledge of design is active: it is purposive. Understanding the nature of design is exposed by asking the following questions. (1) What is its purpose(s)? (2) What is its structure? (3) What are the model cases of it? (4) What are the arguments that explain and evaluate it? There is a communality with the many other heuristics that try to explain problem solving.

Perkins argued that these four design questions are strategic knowledge. They lead to an understanding of principles. He thought that these design questions should be used to convey knowledge. Implicit in this view is the idea that at times the barriers between the knowledge areas (or disciplines) get broken down whether the teacher (philosopher) likes it or not.

Many problems faced by engineers (designers) cannot be solved without a seamless flow of information from different areas of knowledge. The process that produces a product (engineering) requires a technology of organisation for its production. When taken to include "ideas" Engineering and Technological Literacy are necessarily transdisciplinary [12], but to use Whitehead's term each has its own "style" or "way of thinking."

Those who advocate the teaching of design in engineering argue that the solution of engineering problems requires knowledge from several disciplines, and that for the purpose of education, knowledge is integrated by the project method, in addition to appropriately structured scaffolds in the associated curriculum. This could equally be applied to problems in all parts of the knowledge spectrum.

It should be easy to imagine the design of a problem-based course in which the problems are designed to focus on a specific topic in a subject (e.g., economics), that is based on the reality of the day, and that requires the student to systematically find the information necessary for the solution of the problem. In these days of e learning that should be easy for the motivated student. In a problem-based course of this kind, that is learning by doing, the tutor has a significant role in providing quick feedback because in learning by doing it is easy to arrive at faulty conclusions. In Perkins' view the model in Exhibit 1.2 would be re-titled "Design."

The question that is faced by any course leader is, how much knowledge in each of the specialisms is necessary for a person to undertake a transdisciplinary activity? The question might also be put this way—are there alternative ways of organising knowledge that will achieve the same end? Or, what is the necessary content and skills in the disciplinary areas that a person requires to be engineering and technologically literate? A search of the literature on the aims of education reveals Whitehead's theory of rhythm in learning as a possible solution to the problem.

7.5   WHITEHEAD'S RHYTHMIC THEORY OF LEARNING

Whitehead's stage theory is summarized in Exhibit 7.1. The first stage of romance is necessarily one of transdisciplinarity because it is a stage of exploration, a stage of discovery. So too is the final stage of generalisation (synthesis). But between them is precision. It is here that the language, which is the "style" of a particular subject, is learnt; and the interest found in the stage of romance turned into a search for expertise. Whitehead does not expect the stage of romance to be one that is simply a collection of "scraps of information." In his lecture on the aims of education to mathematics teachers he said, "Culture is activity of thought, and receptiveness to beauty and humane feeling. Scraps of information have nothing to do with it. A merely well informed man is the most useless bore on God's earth. What we should aim at producing are [is] persons [men] who possess both culture and expert knowledge in some special direction.

Stage 1: Romance

The stage of first apprehension (a stage of ferment). Education must essentially be a setting in order of a ferment already stirring in the mind: you cannot educate the mind *in vacuo*. In our conception of education we tend to confine it to the second stage of the cycle, namely precision [ ] In this stage knowledge is not dominated by systematic procedure [ ] Romantic emotion is essentially the excitement consequent on the transition from bare facts to first realizations of the import of their unexplored relationships.

Stage 2: Precision

The stage of romance-width of relationship is subordinated to exactness of formulation. It is the stage of grammar, the grammar of language, and the grammar of science. It proceeds by forcing on the students' acceptance a given way of analyzing the facts, bit by bit. New facts are added but they are the facts which fit into the analysis.

Stage 3: Generalization

Hegel's stage of synthesis. A return to romanticism with the added advantage of classified ideas and relevant technique.

Exhibit 7.1: Whitehead's theory of rhythm in the educational process. The stages of mental growth and the nature of education. A summary of pages 27–30 of the essay on "The Rhythm of Education."

Their expert knowledge will give them ground to start from, and their culture will lead them as deep as philosophy and as high as art" [13, p. 1]. Education is then "the acquisition of the art of utilisation of knowledge" [14, p. 6]. Looked at from the perspective of Whitehead's formal philosophy engineering and technology are creative activities. The stage of "romance" is not only one of discovery but of creative exploration [15]. It is a view that fits well with what an engineer seeks to do.

Whitehead argued that education should be a continual repetition of such cycles. While he uses the term stage it is not used in the same sense as Piaget whose stages were relatively fixed. Nevertheless it is useful to think of both long cycles, shorter and very short cycles within them. For example primary (elementary) education is essentially (or should be) a stage of romance. Whitehead praised the Montessori approach to early childhood education. But he also pointed out that an infant's first stage of precision is the mastering of spoken language. This view is consistent with those who believe that young children are capable of handling the "big" ideas of philosophy, and should be taught philosophy in their own terms (language systems) [16].

Similarly, post-primary (high school) education is primarily a stage of precision which is continued through university to be completed by the stage of generalisation. Precision is about the grammar of subjects (disciplines). The speed at which students' accomplish these cycles in a subject is very much a function of their motivation and the way in which subjects are taught.

But for Whitehead education "must essentially be a setting in order of a ferment already stirring in the mind: you cannot educate the mind in vacuo. In our conception of education we tend to confine it to the second stage of the cycle of precision. But we cannot so limit our task without misconceiving the whole task. We are concerned alike with the ferment, with the acquirement of precision, and with the subsequent fruition" [17, p. 29].

It is a criticism of higher education that it confines itself to analysis (precision). It takes no cognisance of the need to motivate students at the beginning of their studies, that is, to create a stage of romance; neither does it necessarily advance them through a stage of generalisation. To be fair, final year project work should go some way to achieving this goal.

Students who are faced with studying a new subject should be allowed to explore the range of the subject. Romance is clearly a stage that should provide a variety of topics with some overall focus (integration) in mind that will motivate a student to want to be precise in a variety of topics. It is not without significance that several engineering educators have in recent years called on their colleagues to take note of what happens in primary (elementary) teaching [18].

It has been pointed out that often engineering students find little connection between the different knowledge units that make up engineering courses. Consequently, there have been some attempts to give engineering courses greater coherence [19]. Whitehead takes the view that learning is a search for coherence which comes about in the stage of generalisation in each rhythmic cycle. There is no doubt that the interface between engineering and society in technology brings about complex problems that demand higher level thinking skills in analysis, synthesis and judgment. In an appropriately designed course the student will be exposed to the excitement of new possibilities and through grammar develop the skill of precision (analysis), and finally through the romance of new and complex sometimes fuzzy problems develop skill in synthesis and judgement. Such a course would bring a student to the higher stages of learning that theorists like Perry, King, and Kitchener believe are not accomplished in higher education [20].

7.6   THE GOALS OF A STAGE OF "ROMANCE" IN A PROGRAM FOR ENGINEERING AND TECHNOLOGICAL LITERACY

The first goal is motivation. Motivation is seldom spelt out as an aim of education yet it is not an un-trivial aim especially where unusual courses, that is, those that step outside the plausibility of the perceivers, are concerned [21]. Currently, programs of engineering and technological literacy (long or short) seem to fit this category. For them to be successful they have to send a message back to the student body that they are interesting, entertaining and worth learning. The key questions for the tutor and curriculum designer are "how do I motivate students through my teaching?" (i.e., "what instructional strategies are most likely to motivate the students?"); "what do I know about the students that will help me motivate them?" and, "am I likely to motivate them with curriculum structures as they are presently organised?" (By structure is meant the linear organisation of the timetable into subjects.) Motivation, instruction, and learning are intimately linked.

The second goal is the exploration of different ways of knowing and learning. It is not at all obvious that entering students will see that it is necessary for them to bridge the gap between the "liberal" and the "vocational." They will have been schooled in educational systems that are classified by subjects and where the distinctions between them are emphasised and therefore, between liberal and professional (vocational) knowledge rather than the seamless pattern to which they belong. For this reason students should be invited to explore different ways of conceiving knowledge including their own, and how it may be re-structured in order that they may use it in specialist study. Related to this is the need to understand how we learn and how we develop the reflective capacity that is indicative of higher order thinking. One of the major advantages of incorporating the fine arts into liberal education is that it forces on the learner an appreciation that there are many ways of thinking about objects in the real world such as connoisseurship.

Elliot Eisner of Stanford University supported this argument with the suggestion that teachers should apply the skills of the art critic to understanding their own performance "the consequence of using educational criticism to perceive educational objects and events is the development of educational connoisseurship. As one learns how to look at educational phenomena, as one sees using stock responses to educational situations and develops habits of perceptual exploration, the ability to experience qualities and their relationships increase. This phenomenon occurs in virtually every arena in which connoisseurship has developed. The orchid grower learns to look at orchids in a way that expands his or her perception of their qualities. The makers of cabinets pay special attention to finish, to types of wood and grains, to forms of joining, to the treatment of edges. The football fan learns how to look at plays, defence patterns and games strategies. Once one develops a perceptual foothold in an arena of activity-orchid growing, cabinet making, or football watching—the skills used in that arena, one does not need the continual expertise of the critic to negotiate new works, or games or situations. One generalises the skills developed earlier and expands them through further applications" [22]. It would seem be self-evident that connoisseurship is a skill that engineering designers should acquire, moreover it is a skill in which the affective is as important as the cognitive. It is a reflective activity that engages the emotions as Macmurray would have it.

"When something of which we are aware attracts our attention so that we stop to contemplate it, we really see it for the first time. We isolate it from its surroundings. Our eyes search it systematically in detail, and discover a hundred things in it that we had overlooked, and these are held together in their intrinsic and particular relations to one another and to the whole which they constitute and to the whole which they constitute. The visual image ceases to be merely schematic and moves towards completeness. This surely the way in which we know an object" [...] [23], and this surely an enlargement of mind which Newman considers to be the goal of university education (see Section 1.8), a cyclic endeavour of romance, precision, and generalisation, oft repeated [24]. It begins with perception and the factors that influence perception.

Apart from the value of understanding how our learning styles influence the way we learn and our responses to different kinds of instruction our perceptions also influence our learning. Perception is an over-arching concept that plays an important part in the way we relate to each other in the workplace, social settings, and the classrooms [25, Chapter 2] [11]. As Bucciarelli pointed out, very often the problems teachers have in seeing the way their students understand a particular problem is because they have not learnt to speak the same language [26, p. 92]. Engineers who are used as expert witnesses have to learn how lawyers use evidence, and what their role is in giving evidence [27, p. 46]. Each of the dimensions of engineering and technology is a different "style" or way of thinking—a different "language."

In science and engineering there have been many studies that show students often have misconceptions of the principles that are to be understood. They have led to new epistemologies such as constructivism in attempts to show why this happens and how the problem can be averted through different approaches to instruction [28, p. 57 ff]. Examination of the arguments for (realism) and against constructivism may provide a basis for students to examine their own epistemologies and values.

The third goal is the exploration of one's personal value system. The base of all engineering and technological activity is the value system we hold. Our beliefs and attitudes drive our personal and working behaviours. The person who is engineering and technologically literate will be grounded in a well thought out ethic. One way of arriving at an ethical position might be to examine the constructivist/realist philosophies in their response to the fundamental issues of ethics [29]. Another way might be to examine theories of moral development such as Kohlberg's [30] and how they might inform self-development (the fourth goal of the stage of "romance") on the one hand, and on the other hand, the concept of moral autonomy in engineering [31].

The fourth goal is to provide for personal development. Whitehead's stage theory is clearly related to his view that "*the valuable intellectual development is self-development*" [32, p. 1]. The teaching strategies we choose can enhance or impede development. As has been argued, most education systems and teaching emphasise cognitive development at the expense of the affective even though it is well understood that in life individuals are expected to work in teams, and that the effectiveness with which teams function is dependent as much on the emotional intelligences of their participants as it is on their cognitive ability. The argument here is, however, that it (development) goes on throughout the whole of life, and that each transition, primarily a change in work and/or personal (family) circumstances is a stage that is accompanied by new insights, and as such, is a stage in development. The idea that intellectual development is self-development commands some assent but it needs to be unravelled further. Clearly, there are two quite different dimensions at issue. There is personal development and there is development in engineering and technological literacy. Are they separate or do they live together? In either case the peak of development is the reflective capacity with which it endows an individual. As indicated above Macmurray pointed out that this must embrace the intellectual and the emotional: both are activities of knowing [33].

A criticism of engineering education, and indeed other subjects within higher education, is that they concentrate on the intellectual at the expense of the emotional. Within recent years there has been recognition by industry that it needs individuals who have a balanced emotional intelligence, that is, it has assigned significance to the affective dimension of human behaviour. Students might be asked to discuss the question, "Given that our actions so often hurt the feelings of others should our understanding of 'reason' embrace feeling and action?" Asking this, or similar questions, invites the students (us) to consider whether thinking and acting, emotion and reason, and freedom and responsibility are opposites? [34, p. 217].

The fifth goal is to provide practical experience in the art and science of engineering, that is the experience of designing and making things. Engineering is an inherently practical activity. It embraces design, investigation, and the making of things. They add skills without which any program of liberal education is incomplete. For example, a traditional academic curriculum mostly neglects the spatial abilities that are important in design and scientific thinking.

It is clear that a stage of romance that is directed towards the attainment of these goals will necessarily draw students through the cycles requiring some precision and some generalisation appropriate to the student's knowledge at the time of their search for understanding. In Bruner's terms it is the first stage of the spiral in a student's personal curriculum [35]. It follows that the provision for learning in a stage of romance has to include activities that will help develop a student's reflective capacity both with respect to himself or herself as a person, or a person engaged in the activities of engineering and technological literacy. And that is the foundation for bridging the gap between the "liberal" and the "vocational."

This is to run against the trend in some countries of placing students in vocational studies earlier. Or, that of redirecting some potential university applicants away from degree programmes that will not produce sufficient earnings for them to be able to pay back the loans they will have to take out.

The programme described here reconciles the vocational with the practical. The contention here, is that it is better to lengthen the period of general education within the context of the epistemology of liberal education as proposed by Newman but in which technology is a necessary component. The research of Eric Hanushek and his colleagues lends support to this view as does the sociological analysis of the semi-professions by T. H. Marshall [36]. Hanushek writes that "policy proposals promoting vocational education focus on the school-to-work transition. But with technological change, gains in youth employment may be offset by less adaptability and diminished employment later in life" [37]. Hanushek and his colleagues found strong robust support for such a trade-off among 11 countries, more especially in those that emphasised apprenticeship programs for that age group. These included Germany whose dual system of education and training many in the UK would like to model.

7.7 THE STAGE OF ROMANCE AND CONTENT

Whitehead has much to say about the organisation of the stage of "romance" but has little to say about content except that a person must be able to explore or better still venture into all the areas of knowledge that contribute, in this case, to engineering and technological literacy. While the teacher should determine what should be learnt the traditional methods of the stage of "precision" will not achieve "romance." Methods more akin to those used in primary (elementary) schools are better tuned for its accomplishment, e.g., projects and case studies. Whitehead attributed the success of the Montessori system to the dominance of romance in the programme [38, p. 62] but as has been shown project work and case studies and methods like debating also require the completion of the other stages of the cycle. It is to quote Edmund Holmes *the path to realisation*" [39, p. 66]. It provides the initial basis for insight into the field of human inquiry and human opinion that is engineering and technological literacy [40, p. x]. It is an adventure into the cycle of knowledge, and in this case, into the different domains of engineering.

The concern is with the engineer's act of understanding *per se* on the one hand, and on the other hand with society's understanding of the products of the processes of engineering-technology. Good engineering occurs when these two understandings merge (Alan Cheville, personal communication). Whatever teaching method is selected, will also be in the service of that end. As conceived here the stage of romance for a program in engineering and technological literacy is an all embracing intensive program of activities supported by independently obtained knowledge from the disciplines (just in time knowledge) that enable a student to discover the properties of engineering and technological literacy that are derived from key concepts representative of the different knowledge dimensions, and the concepts that provide bridging between them.

7.8 CONCLUSION

Currently, the language of Whitehead's stage of "precision" is the primary goal of university courses, and much of the three or four years spent in college is on courses in which it is emphasised. If the predictions discussed in earlier parts of study are in any way correct, the why and what of knowledge reception and skill will change radically. In the two-year programme envisaged in the model presented, year one would be a stage of romance, and year two a stage of preparatory precision. Both years would be of a whole years' duration excluding holidays and include two periods in industry, one of which might be devoted to teaching. In the case of England this would be equivalent to the traditional three years of study.

In the model of "romance" presented above the focus subject is engineering and technology. The transdisciplinary element of the course introduces the student to philosophy, psychology and theology on one side of the spectrum, and on the other side organisational psychology (see Section 1.5) finance, economics, and sociology (see Section 1.5). But other subjects are studied as well. Transdisciplinarity can be derived from problem-based studies on any subject on which the focus is placed (see note [12]).

In this model the life experiences of individuals take them through all the stages of the model. Both the stages of "romance" and "generalisation" are stages when persons have their greatest potential to be creativity. The formal periods of learning are stages of precision that should also accomplish in individuals "renewal."

The view taken here is that the state should be responsible for the costs of basic higher education. In this model it would be for the first two years. It assumes that sufficient savings are made from shortening the course. One way of looking at the problem is to view basic higher education as an extension of school education. A century ago some children left school at the age of 12 or 13. Now many children remain in school until they are 18. We know from neuro-science that brain plasticity continues well into the twenties so this is not an unreasonable proposition.

Given that many individuals' will have to change jobs on a number of occasions during their lifetime then industrial organisations will have to accept that they share with society, (probably represented by an institution of higher education) responsibility not only for the professional but personal development of the individual. This will require a rethinking of the concept of the "firm" and its role in society.

If individuals are likely to work for much less time than they have done in the past as some forecasters predict, the personal begins to become much more important than the professional. That would be to correct the present imbalance between the professional and the personal for it is the "personal life" that is the driver of all our behaviour.

Taking all these issues into account a model of liberal education applicable to all "permanent" (continuing) higher education based on the educational philosophies of Newman, Macmurray and Whitehead was derived from a re-evaluation of the aims of education that serves as a framework for all four realities, the person, the job, technology and society. Without such a programme the conflicts inherent in the concept of technological literacy cannot be resolved.

7.9   NOTES AND REFERENCES

[1] Taba, H. (1962). *Curriculum Development. Theory and Practice*. New York, Harcourt Press. 116

[2] Miertschein, S. and Willis, C. (2017). Using course maps to enhance navigation of E learning environment. *Proc. Annual Conference of the American Society for Engineering Education*, Paper 2363. 116

[3] Steif, P. (2004). An articulation of the concepts and skills that underlie engineering statics. *ASEE/IEEE Proc. Frontiers in Education Conference*, F1F:5–10. 116

[4] Papadopoulos, C. (2017). Assessing cognitive reasoning and learning in mechanics. *Proc. Annual Conference of American Society for Engineering Education*, Paper 2537. 116

[5] Heywood, J. and Montagu Pollock, H. (1977). *Science for Arts Students. A Case Study in Curriculum Development.* Guildford, Society for Research into Higher Education. Shows how a course was redesigned to enhance the motivation of students. 116, 127

[6] Vincenti, W. G. (1993). *What Engineers Know and How they Know It. Analytical Studies from Aeronautical History.* Baltimore, MD, The Johns Hopkins University Press. 117

[7] Davis, M. (1998). *Thinking Like an Engineer. Studies in the Ethics of a Profession.* Oxford, Oxford University Press. 117

[8] Lienhards, J. (2000). *The Engines of our Ingenuity.* Oxford, Oxford University Press. 117

[9] Perkins, D. N. (1986). *Knowledge as Design.* Hillsdale, NJ, Lawrence Erlbaum. 117

[10] *Ibid.* 117

[11] *Ibid.* 117, 122

[12] Transdisciplinary derives from the need to respond to a single complex, concrete problem that requires the assistance of several disciplines that give a variety of viewpoints to the solution of the problem which is not resolvable by a single discipline but requires the synthesis of a number of solutions. This definition has its origins in a 1973 OECD document which is summarised in (a) Heywood, J. (2005). *Engineering Education. A Review of Research and Development in Curriculum and Instruction.* Hoboken, NJ, Wiley/IEEE. For a discussion of various models of interdisciplinarity see (b) Fogarty, R. (1993). *Integrating the Curriculum.* Pallatine IL, IRI/Sky Publ. 118, 125

[13] Whitehead, A. N. (1932). *The Aims of Education and other Essays.* Fourth impression with forward by Lord Lindsay of Birker. London, Ernest Benn. 119, 126, 127

[14] *Ibid.* 119

[15] I have interpreted Whitehead's major concept of creativity to fit this argument, but I think he would have agreed. For Whitehead every concrete entity is an individualization of the universal creative force, (see p. 268 of Lowe, V. (1990). *Alfred North Whitehead. The Man and his Work*, vol. II, Baltimore, MD, The Johns Hopkins University Press). 119, 128

[16] Lipman, M. and Sharp, A. M. (Eds.) (1978). *Growing up with Philosophy.* Philadelphia, PA, Temple University Press. See also Matthews, G. B. (1980). *Philosophy and the Young Child.* Cambridge, MA, Harvard University Press. 119

[17] *Loc. cit.* Ref. [13]. 120

[18] Crynes, B. L. and Crynes, D. A. (1997). They already do it: Common practices in primary education that engineering education should use. *Proc. Frontiers in Education Conference*, 3:12–19. 120

[19] Culver, R. S. and Hackos, J. T. (1982). Perry's model of intellectual development. *Engineering Education*, 73(2):221–226. 120

[20] (a) Perry, W. B. (1970). *Forms of Intellectual and Ethical Development in the College Years*. New York, Holt, Rienhart and Winston. (b) King, P. M. and Kitchener, K. S. (1992). *Developing Reflective Practice*. San Francisco, CA, Jossey Bass. 120

[21] *Loc. cit.* Ref. [5]. Shows how a course was redesigned to enhance the motivation of students. 120

[22] Eisner, E. W. (1979). *The Educational Imagination; On the Design and Evaluation of School Programs*. New York, Collier MacMillan. 121

[23] Macmurray, J. (1957). *The Self as Agent*, pages 200–230, London, Faber and Faber. 121, 127

[24] *Loc. cit.* Ref. [13]. 122

[25] Heywood, J. (2017). *The Human Side of Engineering*. San Rafael, CA, Morgan & Claypool. 122

[26] Bucciarelli, L. L. (2003). *Engineering Philosophy*. Delft, DUP Satellite. 122

[27] Woodson, T. T. (1966). *Introduction to Engineering Design*. New York, McGraw Hill. 122

[28] Heywood, J. (2005). *Engineering Education. Research and Development in Curriculum and Instruction*. Hoboken, NJ, Wiley/IEEE Press. 122

[29] Vardy, P. and Grosch, P. (1994). *The Puzzle of Ethics*, 1st ed., London, Font Paperbacks, Harper Collins. 122

[30] Kohlberg, L. and Turiel, E. (1971). Moral development and moral education. Lesser, G. (Ed.), *Psychology and Educational Practice*. Chicago, IL, Scott-Freeman. 122

[31] Haws, E. D. (2001). Ethics instruction in engineering education: A (mini) meta-analysis. *Journal of Engineering Education*, 90(2):223–231. There have of course been developments since then. See for example Bowen, W. R. (2009). *Engineering Ethics. Outline of an Aspirational Approach*. London, Springer-Verlag. 122

[32] *Loc. cit.* Ref. [13]. 122

[33] *Loc. cit.* Ref. [23, p. 194 ff]. 123

[34] Costello, J. E. (2002). *John Macmurray. A Biography*. Edinburgh, Floris Books. 123

[35] Bruner, J. (1960). *The Process of Education*. New York, Vintage. Bruner considered the curriculum to be dynamic and evolving. In a course based on this principle the curriculum evolves through a process of discovery learning. It is reinforced by a spiral curriculum in which basic concepts are discussed at increasing levels of depth in different contexts and in which there is feedback between the levels (See also Heywood, J. (2018). *Empowering Professional Teaching in Engineering. Sustaining the Scholarship of Teaching*, Chapter 8, San Rafael, CA, Morgan & Claypool). 123

[36] Marshall, T. H. (1939). Professionalism in relation to social structure and policy, reprinted in Marshall, T. H. (1963). *Sociology at the Crossroads and Other Essays*. London, Heinemann. 123

In relation to the growing number of semi-professional institutions in the UK he wrote "an organized profession admits recruits by an impartial test of their knowledge and ability. In theory they are selected on merit, but it is merit of a particular kind which usually must be developed and displayed in a particular, prescribed way. A narrow road leads into the profession through certain educational institutions. How far this favours social mobility depends on whether those institutions are open to the masses, so that merit can win recognition in all classes. Granted that broadening of the educational ladder typical of modern democracies, the system of official examination is more favourable to mobility than one arbitrary appointment or casual promotion. But the chance to move comes early, during schools days. Once it has been missed and a career has been started at a non-professional level, the whole system of formal qualifications makes movement at a later stage well-nigh impossible."

"There is another point. In the church or the army, in law, or medicine, a man at the head of his profession is on top of the world. He admits no superiors. But many of these new semi-professions are really subordinate grades placed in the middle of the hierarchy of modern business organization. The educational ladder leads into them but there is no ladder leading out."

[37] Hanushek, E. A., Schwerdt, G., Woessmann, L., and Zhang, L. (2017). General education, vocational education, and labor market outcomes over the lifecycle. *Journal of Human Resources*. 52(1):48–87. 123

[38] See Ref. [15]. 124

[39] *Ibid*, p. 66. 124

[40] Lonergan, B. J. F. (1958). *Insight. A Study of Human Understanding*. London, Darton, Longman and Todd. 124

[41] Zeidler, D. L. (2014). STEM Education: A deficit framework for the twenty first century? A sociocultural socio scientific response. *Cultural Studies of Science Education*, 2(9).

This model comes close to the criteria discussed in this paper. It derives from the need to respond to deficit thinking in STEM-based initiatives.

CHAPTER 8

Postscript

In a paper of the 1960 summer institute on effective teaching for young engineers the leading authority on the curriculum at the time, Ralph Tyler, in a lecture on "conducting classes to optimise learning," published some extracts from his publications the first of which, was on the construction of the curriculum. He shows just how difficult it is to determine the objectives of the curriculum. In this study a distinction is made between aims and objectives. Aims are general statements whereas objectives, now commonly called outcomes, are statements of expected learning behaviour. They derive from aims, and it is aims to which we are often emotionally attached. For example we talk about motivating students. But, there are aims that relate more generally to the economy, society and the individual. The reader will have observed the position which is taken and supported in this study is that public policy has focused on the economy at the expense of the individual, and that impacts on decisions about what the curriculum should be. The philosophy that drives public policy in the Anglo-Saxon world is utilitarianism implemented through quantified scientific management.

It will also be clear that the curriculum designer is bombarded with numerous objectives and some of which are contradictory. Ralph Tyler argued that because there could be long lists of objectives there was a need to reduce them to those that are significant eliminating contradictions between. He called the process of doing this "screening" which is accomplished in two stages.

First, the educational and social philosophy of the institution and, discipline (in this case-engineering), determine the primary aims, they also clarify terms, and avoid contradictions. Second, the selected objectives are compared with what is known from the psychology of learning. This necessarily has to take into account instructional psychology.

If a broad interpretation is made of social philosophy then aims relating to society and the economy would be taken into account.

The whole process leads to a model of the curriculum process that approximates to the diagram in Exhibit 2.1 which although non-linear does not illustrate the complexity of the activity. It was understood in the 1960s that assessment was integral to the process. It was also understood that assessment teaching and learning had to be aligned although it was not called that, then. This volume was intended to illustrate the major aspects of this process. Covid has brought to a head a ferment that had already begun that was related to the funding of higher education by student fees. It was soon apparent in the UK that the system of loans was dysfunctional, and that attempts to use the system to persuade students to take degrees that would enable them to repay their loans was unworkable. Covid exacerbated matters both in the UK and the U.S.: the supposed benefits of on-line learning did not accrue to the extent that students might resist its use in normal courses where it could be of considerable benefit. Thus, there are calls for change and enquiries into university education. In the UK, *The Times* newspaper has set up its own commission of inquiry.

At the same time there appears to be a movement to return to normality, that is the situation as it was prior to Covid. It will be very difficult to change the system unless those who could change the system are prepared to look beyond and outside their boxes.

Beyond is meant the future and the challenge to resist the temptation to assume that advances in machine learning will not put people out of work but simply displace them. Alternatively, there is, a belief that the green revolution will bring about a massive expansion in the workforce associated to the construction and installation of renewables. This could turn out to be fewer than is imagined. Whatever happens, a large number of people are destined to become at least temporarily, perhaps permanently employed.

What role has education to play in those circumstances? Imagine further that a number of jobs currently done by the middle classes is greatly reduced with a corresponding change in the class structure, what will this mean for educational structure?

Perhaps T. H. Marshall will turn out to be correct, and as competition is further transferred to the school, there will be a demand for the lengthening of compulsory education in such a way that creates a new middle education. Will we be prepared to pay for an extension of that education until, say, a person is twenty?

These are matters for another occasion but this text is a warning that change is likely to be very difficult. For example, it was clear in the nineteen-sixties, in the UK, that there was an opportunity to develop an alternative curricular but these were not taken because of the prestige that a degree offered the Colleges of Advanced Technology. Whether or not the recently established college by the entrepreneur James Dyson that has the ability to award degrees in engineering will influence the rest of the system to forge close links with industry remains to be seen. But, like it or not, status is a powerful factor influencing choice, and the graduate premium is perceived to be better than that of vocational certificates.

Apart from status academic views about how students learn are deeply entrenched. Many consider that the mind is a *tabula rasa*. Few consider that the mind is the most precious instrument we have, or that perhaps the academic environment might be injuring many minds. This study shows that there are other epistemologies. Similarly, policy makes do not seem to be able to stand outside the box and consider alternative ways of student finance other than loans.

This reluctance spills over into educational research. It cannot be said that engineering education research, and there is much of it has had much influence on the system. For example, little notice has been taken of those who have responded to adult developmental psychology, and taken the opportunity to design curricular that are responsive to the needs of cognitive and personal development. To put it more directly, "how many engineering academics want to know what engineers do in their employments so that the curriculum can be adjusted?" As it is two quite different views of what engineering is are to be found in academies and industry.

Those who study the implementation of change show that while it is necessary to get ideas into the system, it is exceptionally difficult. This study is intended to start discussion. Its big idea is the fact that we are prepared to go along with a philosophy that considers the output of the universities to be cogs at the expense of the personal. To put it in another way, universities have very little interest in learning.

APPENDIX A

An Experimental Curriculum

A.1   INTRODUCTION

In the mid-1970s, the Minister for Education in Ireland approved a project that would allow a few schools to develop a transition year (TY) between the junior cycle of post-primary education when students take a public examination called the Junior Certificate (15+ years) and the first year of the two year programme for the Leaving Certificate (17+ years). The idea was that students should be freed from their studies for these public examinations in order that they should undertake studies that would help their personal and career development. They would continue with some traditional studies but the focus of the year would be on non-traditional studies and include work and community experience. By 1986 the Curriculum and Examinations Board had published the list of outcomes (skills and competencies) shown in Exhibit A.1. Since then various Ministers have increased the numbers of schools allowed to offer the scheme and now large numbers of students participate. There is some evidence that those who participate in the transition year perform better in their Leaving Certificate examinations. Within certain limits of approval the schools are largely free to develop their own schemes. It was therefore an ideal scheme with which to experiment with the curriculum and the Christian Brothers who owned a large number of secondary schools (12–18 years) took advantage of this opportunity. One of their endeavours in association with this writer and other colleagues in Trinity College Dublin was to develop a program for technology. Although they tested some of the components of the programme, for political reasons, it was not completed. The program design was based on Whitehead's rhythmic theory of learning and motivation (Exhibit A.2).

A.2   AN EXPERIMENTAL CURRICULUM

Clearly there has to be some closed instruction in a stage of romance, this may be taken to include formal reading. But the spirit of the stage is that of open-ended inquiry as a prelude to precision. In today's jargon the principal strategies induce active learning. It is not confined to anyone method although it is clear that if romance is to be achieved project- (mini and major) and problem-based learning will have a major role to play. Some activities will necessarily be structured by some form of cooperative learning, and the philosophical and legal dimensions might be enhanced by debates. It is a matter of curriculum design to choose the most appropriate strategy for the attainment of the required objective.

- Have been exposed to a broad, varied, and integrated curriculum and developed an informed sense of his/her own talents and preferences in general educational and vocational matters (*transition skills*)
- Have developed significantly the basic skills of literacy and numeracy and oracy. (It is assumed that most students will have developed these skills before the end of the junior cycle, but specific reinforcement for some will be needed through TYO.) (*literacy and numeracy skills*)
- Have developed confidence in the unrehearsed application of these skills in a variety of common social situations (*adaptability*)
- Have experienced as an individual or part of a group, a range of activities which involve formal and informal contacts with adults outside of the school context (*social skills*)
- Have developed confidence in the process of decision making, including the ability to seek out sources of support and aid in specific areas (*decision making*)
- Have developed a range of transferable thinking skills, study skills and other vocational skills (*learning skills*)
- Have experienced a range of activities for which the student was primarily responsible in terms of planning, implementation, accountability, and evaluation, either as an individual or a partner in a group (*problem solving*)
- Have developed appropriate physical and manipulative kills in work and leisure contexts (*physical*)
- Have been helped to foster sensitivity and tolerance to the needs of others personal relationships (*interpersonal/caring*)
- Have been enabled to develop an appropriate set of spiritual, social, and moral values (*faith/morals*)
- Have had opportunities to develop creativity and appreciation of creativity in others (*aesthetic*)
- Have developed responsibility for maintaining a healthy lifestyle, both physical and mental (*health*)
- Have developed an appreciation of the physical and technological environments and their relationship to human needs in general (*environment*)
- Have been given an understanding of the nature and discipline of science and its application to technology through the processes of design and production (*science/technology*)
- Have been introduced to the implications and applications of information technology (*information technology*)
- Have been given an understanding of the nature and discipline of science and its application to technology through the processes of design and production (*science/technology*)
- Have been introduced to the implications and applications of information technology (*information technology*)

Exhibit A.1: The aims of the transition year opportunity (TYO) expressed in terms of the skills and competencies that should be developed in *Planning, Introducing and Developing Transition Year Programmes. Guide Lines for Schools* (1986). Interim Board for Curriculum and Examinations. Dublin. The Board stated that the list was not exhaustive and that new ones may emerge through the experience of schools in offering the curriculum.

**Experimental research program on the transition year
for cost-effective integrated technology**

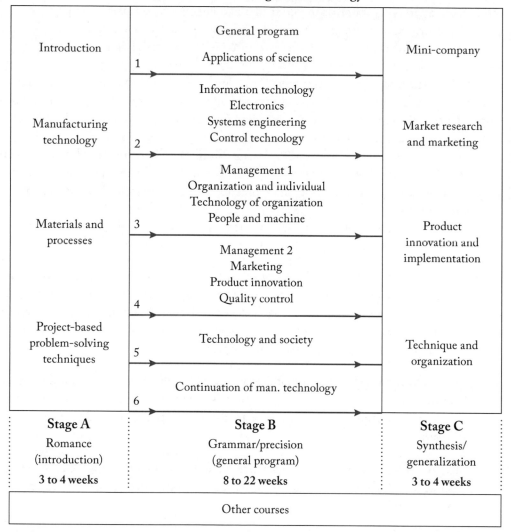

Exhibit A.2: Whitehead's rhythmic process of learning applied to the desing of a programme in technological literacy.

The core activity of this transdisciplinary approach is engineering for it is engineering that produces the technology. A *sine qua non* of this curriculum is that students should experience the process of engineering which is to experience the process of design and manufacture. It was suggested in the earlier paper that students need to know about manufacturing processes and materials and how to use measuring equipment. It was also suggested that this might be achieved through participation in two short but intensive courses in manufacturing technology and technical investigations of the kind that had been reported elsewhere [1, 2].

The former was structured around nine mini projects. It was intended that most of the projects would focus on the processes associated with a material. Two projects were included in order to introduce the students to and its applications in the area of mechanics (structures) and electronics. The final project was intended to allow the students to integrate what they had learnt on the course. In ordered to foster ability in design a section on graphical communication was included early in the course (second exercise). The exercises are shown in Exhibit A.3. Clearly there have been many developments since it was designed that would have to be incorporated in a revised course, as for example—3D printing. Similarly, Sheri Sheppard had not, at that time, described Stanford's course on reverse engineering which would have been an invaluable addition to the course. The course did not include practice and experience with electronic circuitry as for example "breadboards."

The course in technical investigations was also delivered over a three week period with 12 males and 12 females in the age range 16–18. Its objectives were:

1. To develop investigational skills in selecting and isolating parameters to be investigated, selecting and /or devising measurement techniques and equipment, selecting and/or devising procedures, obtaining reliable observations, processing those observations, drawing inferences and estimating the reliability of the inferences, reporting the investigation, recognising and applying optimisation criteria;

2. To develop an appreciation of the ways in which external forces can distort a structure, the ways commonly used structural materials behave under stress; common mechanisms and simple motors and their performance characteristics; the use of pneumatic devises to activate control;

3. To give experience in the use of science to help solve practical problems and to develop an appreciation of the importance of other factors in determining the optimum solution of a given practical problem; and

4. To introduce students to some industrial practices relating to design and quality control.

These objectives were achieved with the aid of 10 mainly structured exercises (e.g., making a crank for turning axles; measurements with wheels and axles; making a reversing switch), and more open-ended investigations into gears, and structures. The final week of the course was devoted to the completion of two projects. The evaluation showed that the fourth objective was not achieved. It was thought that it might have been better achieved if students could have had some work experience. It was found that the students were sufficiently motivated by the "practical," connotations of the program to work well for extended periods. The evaluation led

Mini Project	Purpose(s) – Materials and Processes	Method
	To introduce the students to:	The student is to design and manufacture (except in No 1)
1	(a) Each other (b) Develop teamwork (c) Problem solving methodology	Working in (N) groups in competition to build the tallest tower
2	Computer-assisted machining	A component from a billet
3	Problems in the design and manufacture of acrylic plastic	A simple cassette rack
4	Simple vacuum processes and their potential	A tray insert (e.g., cookery tray)
5	Turning and brazing mild steel	A trophy depicting a sporting activity
6	The principles of aluminum casting	A relief plaque
7	Techniques of cutting and bending sheet metal	A small jewelry box
8 Major project	To enable the students to use their experiences in the mini projects in a more substantial creative activity in which all the skills acquired could be utilized.	A small clock (escapement provided)

Exhibit A.3: The projects in order of their completion. 9 & 10 as planned are not included because they were not completed. Reproduced from Owen, S. and Heywood, J. (1990). Transition technology in Ireland. *International Journal of Technology and Design Education*, 1(1):21–32. The evaluation includes a daily account of what happened on the course. 12 males and 12 females in the age range 16–18 completed the 3-week program which was completed in a custom built laboratory financed by the Irish Christian Brothers.

the authors to conclude that the term "technical investigations" needed to be redefined and in consequence their aims became: The development of:

1. Practical investigative skills.

2. Understanding useful technical devices and commonly used materials.

3. Ability to apply mathematical skills in a useful context.

4. Understanding the uses of science as an aid to living and doing.

5. Appreciate design and quality control as aspects of industrial activity.

Even if the two courses were programmed together there would have been no overarching concept of design, and without that several key aspects of engineering and technological literacy would be missed as, for example, the differences in epistemologies used in engineering and science. Nevertheless, they do highlight the value of such approaches in the achievement of "romance."

A.3   NOTES AND REFERENCES

[1] Owen, S. and Heywood, J. (1990). Transition technology in Ireland: An experimental course. *International Journal of Technology and Design Education*, 1(1):21–32. 138

[2] Kelly, D. T. and Heywood, J. (1996). Alternative approaches to K—12 school technology illustrated by an experimental course in technical investigations. *ASEE/IEEE Proc. Frontiers in Education Conference*, pages 388–391. 138

[3] Carter, G., Heywood, J., and Kelly, D. T. (1986). *A Case Study in Curriculum Assessment. GCE Engineering Science (Advanced)*. Manchester, Roundthorn Publishing.

Author's Biography

JOHN HEYWOOD

John Heywood is a Professorial Fellow Emeritus of Trinity College Dublin–University of Dublin. He was given the best research publication award of the Division for the Professions of the American Educational Research Association for *Engineering Education: Research and Development in the Curriculum and Instruction* in 2006. Recently, he published *The Assessment of Learning in Engineering Education: Practice and Policy*. Previous studies among his 150 publications have included *Learning, Adaptability and Change* and *The Challenge for Education and Industry*, and co-authored *Analysing Jobs*, a study of engineers at work. He is a Fellow of the American Society for Engineering Education, a Fellow of the Institute of Electrical and Electronic Engineers, and an Honorary Fellow of the Institute of Engineers of Ireland. In 2016 he received the Pro Ecclesia et Pontifice award from the Pope for his services to education. In 2018 the TELPhE Division awarded him the best publication award and a meritorious award.